'O' Grade Chemistry
Essential facts and theory
Second edition

R. A. Robertson
Assistant Head Teacher
Rosehall High School
Coatbridge

Revision assisted by

Dr. J. R. Melrose
Principal Teacher of Chemistry
Rosehall High School
Coatbridge

Edward Arnold

© R. A. Robertson 1980

First published 1975
by Edward Arnold (Publishers) Ltd
41 Bedford Square,
London WC1B 3DP

Reprinted 1976, 1977, 1978

Second edition, 1980

ISBN 7131 0477 5

All Rights Reserved. No part of this publication may be reproduced, stored in a retrieval system, or transmitted in any form or by any means, electronic, mechanical, photo-copying, recording or otherwise, without the prior permission of Edward Arnold (Publishers) Ltd.

Acknowledgements

The publishers wish to thank the Scottish Certificate of Education Examination Board for permission to reproduce tables from p. 28, 29, 30, 31, 32, 36 and 40 of the S.C.E. *Mathematical Tables and Science Data Booklet*. These tables appear as an appendix at the end of this book.

British Library Cataloguing in Publication Data

Robertson, Ronald Arthur
　'O' grade chemistry. – 2nd ed.
　1. Chemistry
　I. Title　II. Melrose, J R　III. Series
　540　　QD33

ISBN 0-7131-0477-5

Printed in Great Britain by
WHITSTABLE LITHO LTD.,
Whitstable, Kent

Preface

As long as the examination system exists, there will be a need for a school textbook which covers the 'O' Grade syllabus as briefly as possible, and in a manner suitable for revision purposes. In addition, a textbook of this sort is of great value as a back-up to class work.

This book is not intended to be used as a textbook-workbook; in general, it gives the essential theory and the results which would be obtained from experiments without elaborating on the experimental details. Not only does this condense the book for revision purposes, but it also leaves the class teacher free to devise his own worksheets if he so wishes, and to modify or add experiments in such a way as to reinforce the essential theory covered in the book.

Since a set of data tables is supplied in the 'O' Grade Examination, pupils should be encouraged to use these tables at all times to give them maximum help when writing formulae, ion electron equations, carrying out calculations or deducing information from the Periodic Table.

Often a pupil finds difficulty when writing formulae, because the approach has frequently been to teach the formation of the ionic or covalent bond and, using this information, to write formulae. The pupil was encouraged to decide whether a compound is ionic or covalent and then, obtaining ionic charges or numbers of bonds from the Periodic Table, to write the formula. The process of learning to write formulae can be simplified if the pupil does not have to worry about which type of bond is present. The problem of ionic or covalent character need only be considered when necessary, and certainly after the original 'covalent type' formula has been written. Writing the formula by this method necessitates the pupil obtaining a 'combining number' or 'valency' from the data tables (as indeed, he does anyway when he uses the ionic charge or the number of covalent bonds to write a formula). The use of the word valency went out of fashion in school chemistry because of its connection with the rote learning of the traditional syllabus, but since it is a widely accepted term I have used it in this book.

At the end of each chapter are a number of revision questions which the reader is asked to answer; he may then check his answer from the text. These questions are intended purely for revision purposes, and are not in any way intended to be 'O' Grade type questions.

Although I have a personal preference to teach the work in a somewhat different order from that suggested in the syllabus, I have adhered mainly to the syllabus order, leaving the teacher free to teach the work in the order he prefers.

I would very much like to thank Mr. W. Ross, Advisor in Science in Lanarkshire. Many of his ideas are incorporated in the text. I would also like to thank Mr. W. J. Clelland, Principal Teacher of Chemistry at Bellshill Academy, for his constructive criticism of part of the manuscript, and the Chemistry Staff at my school, Mr. R. Burgess and Mrs. N. Nimmo, for their assistance in testing the content of the book. My thanks must also go to the very helpful secretarial staff at my school who typed the original class notes which formed the basis of this book.

Preface to the second edition

This edition has been brought into line with the revised syllabus of August 1978.

So that the text can be compatible with the first edition, the basic layout has been left unchanged. This means that the order in chapters 1 to 5 is considerably different from the suggested syllabus order. Since we have found the altered approach to atomic structure, formulae and bonding to be beneficial, notes have been included before chapter 1, giving a suggested order of approach to chapters 1 to 5. The data tables have been altered to the new tables supplied in the S.C.E. Exams.

Where it was felt that additional material was necessary (for example, flow diagrams for the industrial preparation of ammonia), this was included as an appendix if it could not be inserted in the text.

Chapters 4.1 and 4.2 (formulae and valency) have been considerably altered, and examples and calculations of empirical formulae (now 4.7) have reverted to a more traditional method. Molarity (now 4.8) has been expanded, with a number of worked examples.

In 6.6, tests for Fe^{2+}(aq) and Fe^{3+}(aq) have been included, and in chapter 7, full 'molecular' type equations are given in addition to ionic equations. Part of the material in 7.4 (the pH scale) has been deleted.

Solubility (8.2) makes use of the new data tables, and at the end of chapter 8, the approach to calculations involving solution concentrations has been changed.

The only other major changes have been in chapter 12, where the fuel gases have been deleted, and a passage inserted on catalytic reforming.

R.A.R.

Contents

Notes to the teacher   viii
1. Atoms and atomic structure   1
2. How atoms combine   9
3. Ions   22
4. Formulae and the mole   25
5. Equations   34
6. Activity and the electrochemical series   39
7. Acids and bases   55
8. Neutralization and salt formation   63
9. Electrolysis of aqueous solutions   74
10. Sulphur and its compounds   78
11. Nitrogen and nitrogen compounds   89
12. Fuels and related substances   101
13. Foodstuffs and related substances   116
14. Macromolecules   133

Appendix: Salt preparations   138
            Industrial preparation of ammonia   139
            Nitric acid manufacture   140
            Data tables   141
Index   148

Notes to the teacher

The use of chapters 1 to 5

In line with ideas of the existing syllabus, chapters 1 to 5 could be approached in a different order to that indicated in the text. Although the order given below shows some minor differences from the syllabus order, you may find the suggested order of value.

1. Atoms and atomic structure.
   As covered in chapter 1.
2. Compounds and formulae.
   2.4 Bond formation and electrons
   2.5 How atoms combine (up to, but not including 'the fate of the shared electrons').
   4.1 to 4.3 Formulae, valency, and writing formulae.
3. Formula mass and the mole.
   4.5, 4.6 and 4.7
   (*Note* Molarity (4.8) should be left until after 8.9, page 71)
4. Equations and their use in calculations.
   5.1, 5.2 and 5.3.

Note At this point, the syllabus moves on to the chemistry of carbon compounds, and if this is the order being followed, section 2.6 from page 15 (the covalent bond) should be done now.

5. How atoms combine.
   2.1, 2.2, 2.3 Conductors and non-conductors.
   2.4 Revision of 'bond formation and electrons'.
   2.5 How atoms combine.
   2.6, 2.7 The covalent and the ionic bonds.
   2.8 Properties of covalent and ionic compounds.
   Chapter 3 — Support for the ionic theory.
   4.4 Writing the structures of ionic and covalent compounds.
6. State equations and ionic equations.
   5.4 and 5.5.

1
Atoms and atomic structure

1.1 Particles

All materials are made up of particles which are continuously in motion and, due to this moving energy (Kinetic Energy), have spaces between them. Some experiments showing this are given below.

1 Diffusion
Liquids and gases gradually mix, regardless of density.

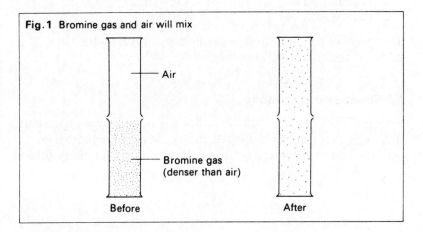

Fig. 1 Bromine gas and air will mix

2 Experiments on mixing
50 cm^3 alcohol + 50 cm^3 water → 98 cm^3 mixture.
 There must, therefore, be spaces between the particles.

3 Smoke – Brownian Movement
Carbon 'specks' in smoke can be seen in the microscope to be jerking about in all directions – they are being bombarded by fast moving but invisible air particles.

1.2 Atoms

All *compounds* are composed of atoms of two or more *elements* joined together. There are just over 100 known elements. The first 92 of them are found naturally and the remainder made artificially.

The elements are composed of tiny particles called *atoms,* and atoms of different elements differ in size and mass.

Symbols for the elements

We use symbols of one or two letters to represent the elements. The first letter is *always* a capital, and when a second letter is used, it is *always* a small letter.

For example:

    Sodium, Na   Oxygen, O
    Carbon, C   Calcium, Ca

The symbols can be found in the Periodic Table in the data tables at the end of this book (on page 147).

1.3 Structure of the Atom

Before 1900, the atom had been thought to be a small, hard ball. However, experiments indicated the existence of a number of particles: the proton, the neutron, the electron, and many others.

The electron

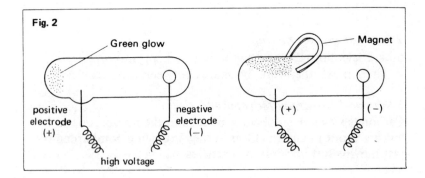

Fig. 2

A cathode-ray tube shows a green, glowing region which can be bent by a magnet, indicating charged particles streaming away from the negative cathode (Fig. 2).
This shows the existence of negatively charged particles, called *electrons*.

The positive nucleus

In 1909, tiny positively charged alpha particles were fired at very thin gold foil (Fig. 3).

Unexpectedly, most of the alpha particles passed straight through and only a few were deflected (about 1 in 10 000). In 1911, Rutherford suggested that most of the mass of the atom was concentrated in a tiny, positive *nucleus* at the centre of the atom, surrounded by negatively charged electrons which take up most of the space.

The positively charged particles in the nucleus were called *protons*.

Also found in the nucleus, are particles with no charge, called *neutrons*.

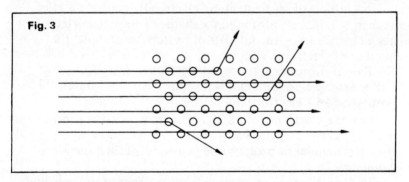

Fig. 3

1.4 Atomic mass scale

0. 000 000 000 000 000 000 000 000 001 67 kg is the mass of a proton. This is obviously too difficult to work with, so a new scale of mass is used — *the atomic mass scale*.

On this scale, the mass of a proton = 1 amu (atomic mass unit), the mass of an electron = about $\frac{1}{2000}$ amu, and the mass of a neutron = 1 amu.

Atomic Structure

Three main particles have been shown to be present:

Fig. 4

		mass	charge
nucleus	proton, p^+	1 amu	+1
	neutron, n	1 amu	0
	electron, e^-	$\frac{1}{2000}$ amu	−1

Arrangement of atoms in the periodic table

The elements are arranged in order of the number of protons which the atoms have. Since the number of protons does not change (the nucleus being protected in the centre of the atom), this number is called the *Atomic Number,* and tells us to which element an atom belongs.

The Atomic Number of an element is the number of protons in the nucleus of one of its atoms.

For example, all atoms of sodium, Na (atomic number = 11) contain 11 protons; all atoms of uranium, U (atomic number = 92) contain 92 protons.

Now, atoms of an element are electrically neutral (that is, uncharged). Since a proton has a charge of 1+ and an electron has a charge of 1−, the number of protons must equal the number of electrons. That is:

Number of protons = Number of electrons.

For example, sodium atoms (atomic No. = 11) contain 11 protons and 11 electrons.

Since the electrons have little or no mass, it is the protons and neutrons in the nucleus which make up the mass of an atom and the total number of protons and neutrons is called the *Mass Number.*

The Mass Number of an atom is the number of protons plus the number of neutrons in its nucleus.

The number of neutrons in an atom

If we know the atomic number and the mass number of an atom, we can work out how many neutrons the atom has. For example:

(a)   Sodium, Atomic Number = 11   ∴ 11 p^+
           Mass Number   = 23

∴ No. of p⁺ + No. of n = 23
i.e. 11 p⁺ +   12 n = 23
(b) Uranium, Atomic Number = 92; Mass Number = 237
∴ the mass of 237 is made up of 92 p⁺ and 145 n.

1.5 The Mass Spectrometer — Determining Mass Number

If a neutral atom has an electron pulled off it, it leaves the atom with a positive charge. A charged atom is called an *ion*. If these positive ions are passed through an electric or magnetic field, they are deflected, the lighter particles being deflected most (Fig. 5).

Particles of different mass strike the detector at different points, and a chart is obtained which shows the mass and proportion of each ion present (Fig. 6).

For example, chlorine gas shows atoms of two different masses to be present.

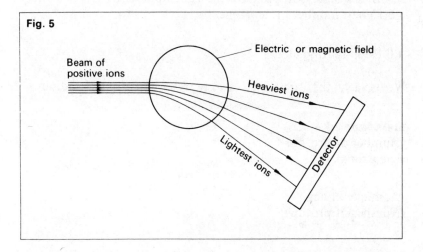

Fig. 5

The position of the lines tells us the mass of the atoms, and the height tells us the proportion present in the sample. Thus the mass spectrometer shows that, although atoms of one element must all have the same number of protons (the same Atomic Number), *they can contain different numbers of neutrons.*

For chlorine, the atomic number is 17. Therefore all chlorine atoms must have 17 protons. This means that the chlorine atom of mass number 35 must have 17 protons + 18 neutrons. The

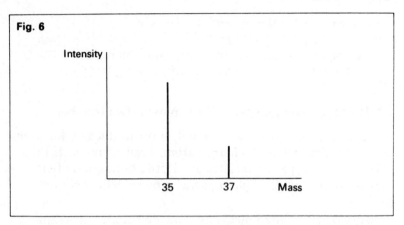

Fig. 6

chlorine atom of mass number 37 must have 17 protons + 20 neutrons.

These different chlorine atoms are called *isotopes* of chlorine. **Isotopes are atoms of an element with different mass numbers. In other words, isotopes have the same number of protons but a different number of neutrons.**

1.6 Representing atoms

We use a symbol such as $^{23}_{11}$Na to represent an atom of sodium.

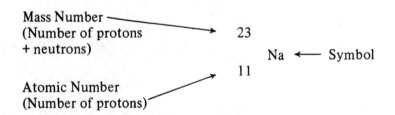

Mass Number (Number of protons + neutrons) → 23

Na ← Symbol

11

Atomic Number (Number of protons)

Relative Atomic Mass (Atomic Weight)

Most elements contain a mixture of isotopes, so it is necessary to know the average mass of atoms of an element. This is the Atomic Mass.

The Atomic Mass of an element is the average mass of an atom of the element, taking into account the proportion of each isotope present.

Atoms and atomic structure

For example, the element *chlorine* has two isotopes, $^{35}_{17}Cl$ (75%) and $^{37}_{17}Cl$ (25%). Since there are more $^{35}_{17}Cl$, the average mass works out at 35.5 amu. That is, the Atomic Mass of Chlorine = 35.5 amu.

1.7 The electrons

The negatively charged electrons are found outside the nucleus. They have energy which prevents them being pulled into the nucleus by the positively charged protons.

The electrons occupy various *energy levels* outside the nucleus, and since they are on the outside, they are responsible for the behaviour of an atom in a chemical reaction.

We can represent the electron arrangement of atoms as in the chart below. Carbon, for example, has two electrons in the first energy level and four in the next.

Electron arrangement of elements 1-20

H 1							He 2
Li 2.1	Be 2.2	B 2.3	C 2.4	N 2.5	O 2.6	F 2.7	Ne 2.8
Na 2.8.1	Mg 2.8.2	Al 2.8.3	Si 2.8.4	P 2.8.5	S 2.8.6	Cl 2.8.7	Ar 2.8.8
K 2.8.8.1	Ca 2.8.8.2						

Elements in the same column have the same number of outer electrons. They behave in a similar manner. The elements in the column with filled energy levels (helium, neon, argon, etc.) are very stable, and are called the *Noble Gases*.

The first energy level is filled with 2 electrons. The second energy level is filled with 8 electrons. The other energy levels are also filled with 8 electrons, but at a later stage this number can be increased.

1.8 Information, and where to find it

The information giving the *Atomic Number, Atomic Mass,* and *Electron Arrangement* of the elements can be obtained in the Data Tables at the end of this book.

Some revision questions

You should try to answer each question and then check your answer by referring to the section indicated in brackets after the question.

1 Copy and complete the following table, listing the three main particles in the atom. (Section 1.4).

Particle	Mass (a.m.u.)	Charge
	1 a.m.u.	
		0
		−1

2 What is meant by: (i) atomic number (ii) mass number (iii) isotopes (iv) atomic mass (atomic weight)? (Section 1.4 to 1.6).

3 Carefully redraw this table and fill in the blanks.

Symbol	Atomic Number	Mass Number	No. of protons	No. of neutrons	No. of electrons
$^{23}_{11}Na$	11	23	11	12	11
	8			8	
			17	18	
		81	35		
				16	16
			24		12

(Section 1.4 to 1.6).

4 Give the electron arrangement for the atoms of atomic number (i) 15, (ii) 7, (iii) 18, (iv) 3, (v) 20. (Section 1.7.)

2

How atoms combine

We can get some idea of how atoms combine if we look at the electrical properties of materials.

2.1 Conductors and non-conductors

Experiments showing the ability of materials to conduct electricity indicate three different types.

(a) *Metallic conductors* – good conductors of electricity in the solid or liquid state. They are not decomposed (broken up) by electricity.

(b) *Electrolytes* – substances which conduct electricity when dissolved in water, or when molten. These substances are decomposed by electricity. Electrolytes are usually solutions of acids, solutions of compounds containing a metal (salt solutions) or melts of compounds containing a metal.

(c) *Non-conductors* – do not appear to conduct electricity whether solid, liquid or in solution. Non-conductors are usually non-metallic elements, or compounds containing non-metallic elements.

Of these three we must now look at *electrolytes* and *non-conductors* in more detail.

2.2 Electrolytes

Let us look at the sort of result we get by passing a current through an electrolyte. (We call this process 'Electrolysis'.)

Electrolyte	Results of electrolysis	
	At negative electrode	At positive electrode
Hydrochloric acid solution	Hydrogen	Chlorine
Copper chloride solution	Copper	Chlorine
Molten lead bromide	Lead	Bromine

Since the hydrogen, copper and lead are always discharged at the negative electrode then we can say that the hydrogen in the hydrochloric acid, the copper in the copper chloride and the lead in the lead bromide must have a positive electric charge (opposite charges attract) and be free to move to the negative electrode.

In the same way the chlorine in the copper chloride and the bromine in the lead bromide must have a negative charge and be free to move to the positive electrode.

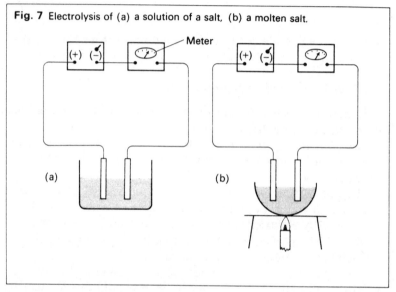

Fig. 7 Electrolysis of (a) a solution of a salt, (b) a molten salt.

We call these charged atoms *ions*.

Since salts conduct electricity in solution or when molten, but not when in the solid state, ions must be held in position in the solid state, and become mobile only when the substance is dissolved in water or melted.

2.3 Non-conductors

These substances do not conduct whether in the solid state (e.g. sugar), the liquid state (e.g. alcohol or molten wax), or in solution (e.g. sugar solution).

This suggests that the particles in non-conductors must be uncharged. We call these uncharged particles *molecules*.

2.4 Bond formation and electrons

For atoms to combine they must collide, and on collision it is their outer electrons which must meet.

How atoms combine

We have already shown the electron arrangement of some atoms. For example, the third row of the Periodic Table consists of:

Na	Mg	Al	Si	P	S	Cl	Ar
2.8.1	2.8.2	2.8.3	2.8.4	2.8.5	2.8.6	2.8.7	2.8.8

Within each energy level, the electrons are found in compartments called *orbitals* (or electron clouds), each orbital (or cloud) being able to hold 2 electrons (an electron pair).

What is an orbital?

An orbital (or electron cloud) is a space occupied by 1 or 2 electrons. Calculations show that the orbitals seem to have a particular shape, and the electrons are moving about so rapidly within these shapes that the orbital can be thought of as a negatively charged cloud.

Orbitals and the energy levels

The *first energy level* has 1 orbital and can hold 2 electrons. The *other energy levels* have 4 orbitals and can hold 8 electrons.

(The third level and those further away from the nucleus can at later stages hold more than 8 electrons, but this can be ignored at present.)

Since it is the electrons in the outer energy levels which are involved in collisions between atoms, we need only consider the outer levels.

When the atoms are combining, the orbitals in the outer energy levels are arranged as far apart from each other as possible, and take up a 3-dimensional tetrahedral shape.

Fig. 8 Tetrahedral arrangement of orbitals

The electrons, if it is possible, occupy separate orbitals. No two electrons occupy the same orbital if there are others vacant.

The outer electron arrangement of the first 20 atoms can therefore be represented as in the chart below.

First Energy Level   H·   He|
                  half-filled orbital   filled orbital
                                          (electron pair)
Other levels:

Group	1	2	3	4	5	6	7	8
Second Level	Li·	Be·	·Ḃ·	·Ċ·	·N̄·	|Ō·	|F̄·	|N̄e|
Third Level	Na·	Mg·	·Al·	·Si·	·P̄·	|S̄·	|C̄l·	|Ār|
Fourth Level	K·	Ca·	etc.					

From the table, the number of half-filled orbitals, or filled orbitals in any particular atom can readily be seen.
For example,

    Na·   1 half-filled orbital.

    ·Ċ·   4 half-filled orbitals.

    |C̄l·   1 half-filled orbital, 3 filled orbitals.

    H·   1 half-filled orbital (remember that the first energy level only has 1 orbital).

The noble gases (helium, neon, argon, etc.) are very stable. When atoms combine, they tend to achieve the electron arrangement of the noble gases which is a full outer energy level (all the orbitals filled).

2.5 How Atoms Combine

All atoms combine in the same way. This combination involves the outer electrons. Before atoms can join together, they must *collide*, and their half-filled orbitals overlap. (That is, their half-filled electron clouds merge.) Fig. 9 shows this happening to two atoms A and B.

The electrons are being shared, with the positive nucleus of

How atoms combine

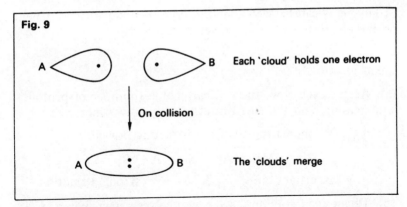

Fig. 9

each atom attracting both electrons and therefore holding the atoms together.

The fate of the shared electrons

There are three possibilities, as shown in Fig. 10.

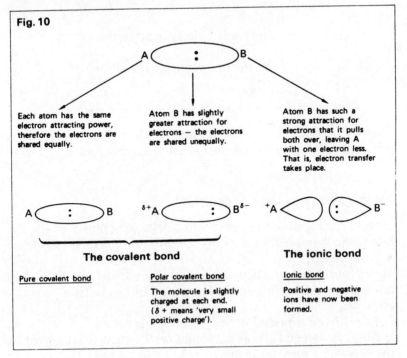

Fig. 10

From this it can be seen that the type of bond formed depends on the electron attracting power of the atoms.

Electron attracting power

(a) Across each row: the nuclear charge (number of protons) is increasing, and the electron attracting power increases.

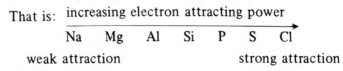

That is:

increasing electron attracting power →

Na   Mg   Al   Si   P   S   Cl

weak attraction   strong attraction

(b) Down each column: each row involves a new energy level, so the distance of the outer electrons from the nucleus increases greatly. Therefore, electron attracting power decreases.

That is: F   Cl   Br   I →

decreasing electron attracting power

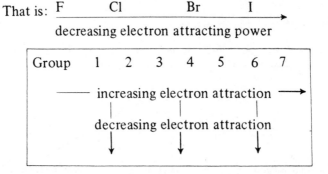

Thus fluorine (Group 7) has the greatest electron attracting power, and caesium (Group 1) has the lowest attracting power.

2.6 Will a bond be ionic or covalent?

If there is a big difference in electron attracting power of the two atoms, the bond will be ionic. If the difference is small, the bond will be covalent.
Ionic bond. Formed when metal atoms (weak electron attracting power) combine with non-metal atoms (strong electron attracting power).
Covalent bond. Formed between non-metal atoms.
   Note: Non-metal atoms are found on the right hand side of the Periodic Table, while metals tend to be on the left.

How atoms combine

The covalent bond

Formation

Atoms have similar electron attracting power (Groups 4, 5, 6, 7, or hydrogen). Atoms collide, and their half-filled orbitals overlap.

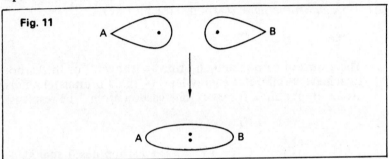

Fig. 11

The electrons remain shared, and both orbitals are filled by this process.

A *covalent bond* has been formed.

Examples

(a) *The chlorine molecule,* Cl_2

    chlorine atom   + chlorine atom   chlorine
  (one half-filled orbital) (one half-filled orbital)   molecule

    $|\overline{Cl}\cdot$ + $\cdot\overline{Cl}|$ ⟶ $|\overline{Cl} - \overline{Cl}|$

                                                                     'shared pair'
                                                                     of electrons

Since we are mainly interested in shared electrons (the bond), we can rewrite this,

         Cl·   +   ·Cl   ⟶   Cl–Cl

(b) *The oxygen molecule,* O_2

The oxygen atom has two half-filled orbitals, and these can overlap with two half-filled orbitals from another oxygen atom, forming a *double bond*.

         O: +   :O   ⟶   O=O

(c) *The nitrogen molecule*, N_2

 $\longrightarrow$ $N\equiv N$

A *triple bond* has been formed.

(d) *Tetrachloromethane* (carbon tetrachloride), CCl_4

·Ċ·   ·Cl

The atoms of carbon and chlorine are shown. For the carbon to achieve a full outer energy level (4 filled orbitals) 4 chlorine atoms are required for every one carbon atom. The result is:

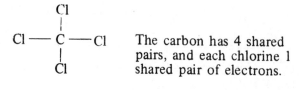

The carbon has 4 shared pairs, and each chlorine 1 shared pair of electrons.

Polar molecules

Since atoms of different elements have different electron attracting powers, there will be unequal sharing of electrons in most cases. In other words, the bond will be *polar* (it will have very small electric charges at each end).

Examples

(a) *The hydrogen chloride molecule*, HCl
    hydrogen atom   +   chlorine atom
    (one half-filled orbital)   (one half-filled orbital)
    H·   ↓   ·Cl

hydrogen chloride
H—Cl
'shared pair' of electrons.

Since chlorine has a stronger attraction for electrons than hydrogen, the bond will be polar:
$$^{\delta+}H - Cl^{\delta-}$$

How atoms combine

(b) *Water molecule, H_2O*

$$H\cdot \qquad :\!O$$

The oxygen atom has 2 half-filled orbitals and requires 2 shared pairs, whereas the hydrogen atom has only 1 half-filled orbital and requires one shared pair.
If the oxygen is to have its 2 shared pairs, 2 hydrogen atoms are required.

$$O: \quad \begin{matrix}\cdot H \\ \\ \cdot H\end{matrix} \quad \longrightarrow \quad O\!\!\begin{matrix}\diagup H \\ \\ \diagdown H\end{matrix}$$

Because the oxygen has a stronger electron attraction, the water molecule will be polar:

$$\delta-O\!\!\begin{matrix}\diagup H^{\delta+} \\ \\ \diagdown H_{\delta+}\end{matrix}$$

The shapes of molecules

Since combining atoms tend to arrange their 4 orbitals in a 3-dimensional tetrahedral structure, we should be able to predict the shape of many molecules. Some examples are given below.

(a) *Tetrachloromethane* (carbon tetrachloride) CCl_4

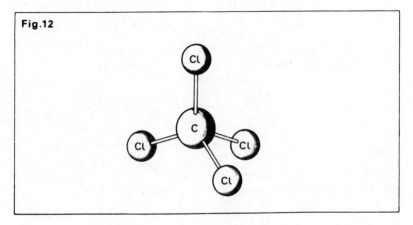

Fig.12

(b) *Water*

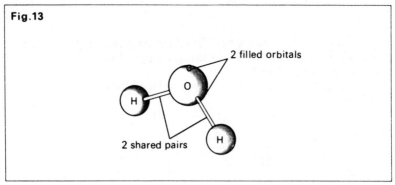

A water molecule should have the shape

2.7 The ionic bond

Formation

Combining atoms have a big difference in electron attracting power. Atoms collide, and their half-filled orbitals overlap.

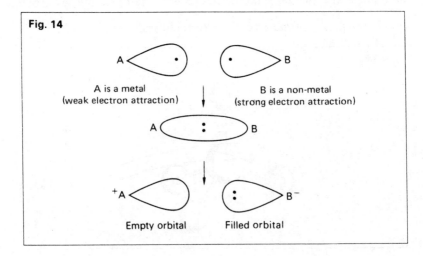

How atoms combine

An electron has been pulled from A completely over to B, and *ions* have been formed. B has gained an electron, becoming negatively charged, while A has lost an electron, becoming positively charged.

Examples

(a) An example of a Group 1 metal with a Group 7 non-metal is sodium chloride.

Sodium, Na, 2.8.1.   Chlorine, Cl, 2.8.7.
1 half-filled orbital.   1 half-filled orbital.
                                    3 filled orbitals.

Each chlorine atom can pull the outer electron from a sodium atom.

```
          1 electron
Na, 2.8.1 ──────────→ Cl, 2.8.7
```

The result is sodium chloride:

Na$^+$, 2.8.   Cl$^-$, 2.8.8.
Complete outer energy   Complete outer energy
level — the electron in the   level — 4 filled orbitals.
orbital of the third energy
level has disappeared.

(b) An example of a Group 2 metal with a Group 7 non-metal is calcium chloride.

Calcium, Ca, 2.8.8.2.   Chlorine, Cl, 2.8.7.
(2 half-filled orbitals)   (1 half-filled orbital)

Each chlorine atom can only pull 1 electron from the calcium, and, since the calcium will have a complete outer energy level if it loses 2 electrons, then 2 chlorine atoms are required for each calcium.

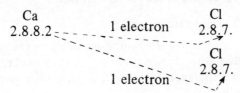

Two electrons are transferred, one to each chlorine atom. The result is calcium chloride, $CaCl_2$

Ca^{2+} 2.8.8. Cl^- 2.8.8.
              Cl^- 2.8.8.

Shapes in ionic compounds

Since ions exist separately, there are no molecules in ionic compounds. For example, a crystal of sodium chloride is composed of a large number of positive and negative ions, held strongly together by electrostatic forces (attraction of opposite charges).

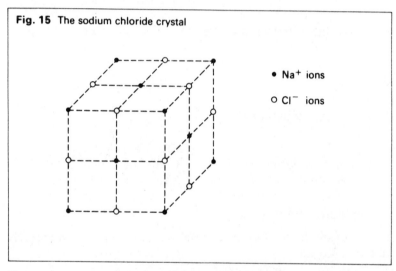

Fig. 15 The sodium chloride crystal

• Na^+ ions
○ Cl^- ions

From this we see that the sodium chloride crystal takes up a cubic shape.

In other ionic compounds the ions are held together in a similar manner. However, they may take up different patterns. In other words, differently shaped crystals can be formed.

2.8 Properties of ionic and covalent compounds

Properties of ionic compounds

1 Because ionic compounds are composed of a large number of positive and negative ions strongly attracting each other, they are solid at room temperature, and tend to have high melting points and boiling points.

How atoms combine

2 When dissolved in water or molten, they become electrolytes. That is, they conduct electricity, and are decomposed in the process.

Properties of covalent compounds

1 The molecules of covalent compounds all exist separately. There is little attraction between them, and they can therefore be easily separated. They tend to have low melting or boiling points. At room temperature, about 20 °C or 293K. they are usually liquids or gases. Those which are solids have a low melting point.

2 Since they have no charged particles, they do not conduct electricity whether solid, liquid, or in solution.

Some revision questions.

You should try to answer each question and then check your answer by referring to the section indicated in brackets after the question.

1 What is an electrolyte? (Section 2.1).
2 (a) Describe what happens to electrons in the formation of
(i) the covalent bond, (ii) the ionic bond.
(b) Draw the three-dimensional shape of the following:
(i) a water molecule, (ii) a molecule of ammonia (NH_3),
(iii) part of the sodium chloride crystal.
(c) Why can we talk about molecules of water, but not about molecules of sodium chloride? (Section 2.4 to 2.7.)

3

Ions

3.1 Evidence for the existence of ions — electrolysis

Experiments in chapter 2 showed that salts conducted electricity when in solution in water, or in the molten state. When in the solid state, no conduction took place.

We suggested that since the metal (e.g. copper in copper chloride solution, and lead in molten lead bromide) was always deposited on the negative electrode (the cathode), then the metal particles in these compounds must be positively charged.

In the same way, we suggested that the non-metal particles in these salts had a negative charge.

On the evidence, it would therefore seem likely that ions exist, the salts conducting when in solution in water or in the molten state when the ions are free to move, but not conducting in the solid state when the ions are held in position.

3.2 Colour of compounds in solution

On looking at the colour of compounds in solution it can be seen that, for example, most potassium compounds are colourless, but potassium chromate is yellow. This suggests that the yellow is caused by the chromate ions. Since other chromate compounds are usually yellow, then it would seem reasonable that chromate ions are yellow.

On looking at other compounds in solution, it can be seen that:
dichromate ions are orange;
copper ions are blue;
nickel ions are green.
However, many ions are colourless (for example, sodium ions, potassium ions, chloride ions, sulphate ions, etc.).

Ions 23

3.3 Seeing ions move

If copper dichromate solution contains positively charged, blue copper ions and negatively charged, orange dichromate ions, then if an experiment is set up in a U-tube as in Fig. 16, we should be able to see the ions move. This is shown by a blue colour developing round the negative electrode, and an orange colour developing round the positive electrode.

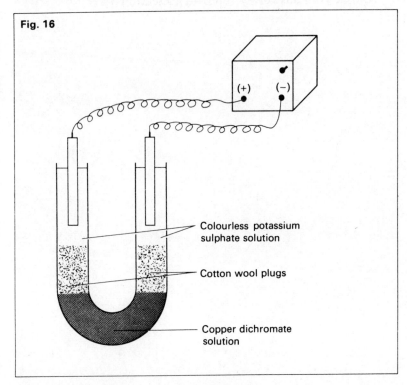

Fig. 16

3.4 Ions containing more than one atom

Some ions are found which contain more than one element; the formula and names of these ions can be found in the data tables at the end of this book.

Example

The formula for the sulphate ion, SO_4^{2-}, means that 1 sulphur and 4 oxygen atoms are bonded covalently to form a group which is only stable when it has gained 2 extra electrons, giving an overall charge of '2−'.

A revision question

How can the electrolysis of copper chloride solution or molten lead bromide indicate that ions exist?
  (Check your answer by referring to section 3.1.)

4

Formulae and the mole

4.1 Writing formulae

A formula tells us what atoms are present in a compound, and the ratio of the atoms to each other.

Since atoms combine by collisions between their half-filled orbitals, then the ability of atoms to join up will depend on how many half-filled orbitals the atom has.

Example – Water
Oxygen has 2 half-filled orbitals, and hydrogen has one:

$$H\cdot \quad \cdot \underset{.}{O} \quad \cdot H$$

The oxygen atom requires 2 shared pairs, while the hydrogen atom can only have one shared pair.

If the oxygen is to have 2 shared pairs, 2 hydrogen atoms are required and the formula is H_2O.

This formula can be confirmed by experiment. When hydrogen combines with oxygen, the ratio H : O is 2 : 1.

4.2 Valency

The number of half-filled orbitals gives us the combining number, or *valency* of an atom.

Obtaining the valency number to write formulae

We can obtain the valency by using information from the data tables on page 143. From the column headed 'Ionic Radii' we can obtain the charge on each ion. The ionic charge and valency are numerically the same.

Element Symbol	Ionic Radii		Ion Charge	Valency
H	H$^-$	0·154	1−	1
He				
Li	Li$^+$	0·068	1+	1
Be	Be^{2+}	0·035	2+	2
B	B^{3+}	0·023	3+	3
C	C^{4+}	0·016	4+	4
N	N^{3-}	0·171	3−	3
O	O^{2-}	0·132	2−	2
F	F$^-$	0·133	1−	1

From Data Tables

If the ionic charge is used to obtain the valency of atoms, then it can also be used to obtain the valency of ions with more than one atom (Data Tables, Page 145).

Examples

from tables	formula	valency
Sulphate SO$_4^{2-}$	SO$_4$	2
Nitrate NO$_3^-$	NO$_3$	1
Ammonium NH$_4^+$	NH$_4$	1

When writing formulae involving these large groups, the large group should be treated as if it were a single 'atom'.

Using the valency, we can now write the formula of many compounds.

Examples

	symbols	valency	simplify valency	formula
sodium chloride	Na Cl	1 1	1 1	NaCl
calcium chloride	Ca Cl	2 1	2 1	CaCl$_2$
magnesium nitride	Mg N	2 3	2 3	Mg$_3$N$_2$

Formulae and the mole 27

	symbols	valency	simplify valency	formula
tetrachloromethane	C Cl	4 1	4 1	CCl_4
carbon dioxide	C O	4 2	2 1	CO_2
sodium sulphate	Na SO_4	1 2	1 2	Na_2SO_4
ammonium carbonate	NH_4 CO_3	1 2	1 2	$(NH_4)_2CO_3$
zinc oxide	Zn O	2 2	1 1	ZnO

4.3 Atoms with more than one valency

Some metal atoms can have more than one valency. In such cases, the valency number is indicated by Roman numerals after the metal.

eg. Copper (I) Oxide, Cu_2O ; Copper (II) oxide, CuO
     Iron (II) oxide, FeO   ; Iron (III) oxide, Fe_2O_3

Practice makes perfect

The only way to master formulae, is to try writing formulae for as many compounds as you can.

4.4 Structure of compounds

Sometimes it is useful to show the structure of compounds and to do this we must decide whether the compound is ionic or covalent.

General rules

Ionic bond	positive ions	metals or NH_4^+ (ammonium ion)
	negative ions	non-metals of groups 5, 6, 7, or non-metal groups such as sulphate (SO_4^{2-}), nitrate (NO_3^-), etc.
Covalent bond	non-metals of groups 4, 5, 6, 7 or H.	

In other words, a compound is ionic if a metal, or the ammonium ion, is present. Otherwise it is covalent.

Ionic formulae

To write an ionic formula, look up the ionic charges in the data tables and write the formula as on the next page.

Compound	formula	ionic formula
sodium chloride	NaCl	Na^+Cl^-
calcium chloride	$CaCl_2$	$Ca^{2+}(Cl^-)_2$
magnesium nitride	Mg_3N_2	$(Mg^{2+})_3(N^{3-})_2$
sodium sulphate	Na_2SO_4	$(Na^+)_2 SO_4^{2-}$
ammonium carbonate	$(NH_4)_2CO_3$	$(NH_4^+)_2 CO_3^{2-}$

For ions with more than one valency, the Roman numerals tell us the ionic charge.

Examples: Copper (I) oxide $Cu_2O, (Cu^+)_2 O^{2-}$
           Iron (III) oxide $Fe_2O_3, (Fe^{3+})_2(O^{2-})_3$

Structural formulae of covalent compounds

We know that:
Number of half-filled orbitals = Valency.

The valency of an atom tells us the number of shared pairs of electrons (bonds) that each atom must have, and having written the formula, we can draw the structure to show the bonds between the atoms.

compound	symbols	valency = no. of bonds	formula	structural formula
hydrogen chloride	H Cl	1 1	HCl	H—Cl
water	H O	1 2	H_2O	O / \\ H H
tetrachloro-methane	C Cl	4 1	CCl_4	Cl | Cl—C—Cl | Cl
carbon dioxide	C O	4 2	CO_2	O=C=O
ammonia	N H	3 1	NH_3	H—N—H | H

Formulae and the mole

Remember that for covalent compounds formed with single bonds, we can draw the shape in 3-dimensions.

4.5 Formula mass and percentage composition

The formula mass (formula weight) is obtained by adding together the masses (weights) of the atoms in the formula.
Copy out and complete the following table.

Compound	Formula	No. of atoms of each element present		Total Mass (Weight) of Atoms	Gram formula mass (weight)
		Atom	No.		
water	H_2O	H	2	2	18 g
		O	1	16	
carbon dioxide	CO_2	C	1	12	44 g
		O	2	32	
ammonium sulphate	$(NH_4)_2SO_4$	N	2	28	
		H	8	8	
		S	1	32	
		O	4	64	
sodium carbonate	Na_2CO_3				
sodium chloride					
tetrachloromethane	CCl_4				
calcium carbonate					

Percentage composition (by mass)

This is best explained by giving some examples.

(a) To find the percentage of hydrogen in water, H_2O.
The formula mass (formula weight) = 18 amu.
That is, in every 18 amu of water 2 amu are hydrogen and 16 amu are oxygen.
Therefore % of hydrogen in water is $\frac{2}{18} \times 100 = 11.1\%$.

(b) To find the percentage of carbon in carbon dioxide, CO_2.
The formula mass = 44 (as above).
Therefore, in every 44 g of CO_2 there are 12 g of carbon.
Therefore % carbon in carbon dioxide is $\frac{12}{44} \times 100 = 27.3\%$.

4.6 The Mole

A mole of a substance is the formula mass expressed in grams. That is, 1 mole = 1 gram formula mass.

The number of 'formulae' in a mole is very large and is in fact about 600 000 000 000 000 000 000 000 (six hundred thousand million million million).

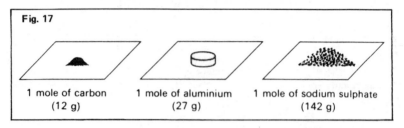

Fig. 17

1 mole of carbon (12 g)   1 mole of aluminium (27 g)   1 mole of sodium sulphate (142 g)

Why do we use the mole?

In theory we could make a molecule of water by reacting 2 atoms of hydrogen and 1 atom of oxygen.

In practice this is impossible, and we must react very large numbers of atoms or molecules together.

We therefore take a formula to mean *one mole* of a material, and can say that, in the case of water (H_2O), 1 *mole* of water (18 g) is formed from 2 *moles* of hydrogen atoms (2 g) and one *mole* of oxygen atoms (16 g).

Also, in the case of sodium sulphate (Na_2SO_4), one *mole* of sodium sulphate contains 2 *moles* of sodium ions and one *mole* of sulphate ions.

4.7 Chemical formulae from experiment (empirical formulae)

We have so far deduced chemical formulae by theory, based on atomic structure. Originally, however, formulae were deduced experimentally by analysis or synthesis, and a knowledge of atomic weights (found from tables).

Experiments can be carried out in class to demonstrate this.

Examples

1. When 20 g of copper (II) oxide was reduced to copper, it was found that 16 g of copper was formed. Find the formula of the copper oxide.

Formulae and the mole

From experiment, 16 g copper combined with 4 g oxygen. We can calculate the formula in the following way:

Elements	Cu	O
Weights combined	16 g	4 g
Atomic Weights	64	16
Weight / Atomic weight = No. moles of atoms	$\frac{16}{64} = 0.25$	$\frac{4}{16} = 0.25$

$$\text{Ratio Cu : O} = 0.25 : 0.25$$
$$= 1 : 1$$
$$\text{Empirical Formula} = CuO$$

2 20.7 g lead were dissolved in nitric acid, and excess sodium iodide solution added. The lead iodide precipitate obtained was filtered, dried and weighed. Its weight was found to be 46.1 g. Find the formula of the lead iodide.

20.7 g lead reacted with $(46.1 - 20.7) = 25.4$ g iodine.

Elements	Pb	I
Weights present	20.7 g	25.4 g
Atomic weights	207	127
Weight / Atomic weight = No. moles of atoms	$\frac{20.7}{207} = 0.1$	$\frac{25.4}{127} = 0.2$

$$\text{Ratio Pb : I} = 0.1 : 0.2$$
$$= 1 : 2$$
$$\text{Empirical Formula} = PbI_2$$

4.8 Molarity

Since reactions are very often carried out in solution in water, we must be able to discuss concentrations of solutions.

A molar solution is a solution containing 1 mole of a substance in 1 litre of solution.
e.g. NaCl. 1 mole weighs $(23 + 35.5)$ g $= 58.5$ g.
Therefore, M NaCl contains 58.5 g in each litre of solution.
2M NaCl contains 117 g in each litre of solution.

Using molarity

Molarity = No. of moles per litre

i.e. Molarity, M = $\dfrac{\text{No. of moles, n}}{\text{Volume in litres, V}}$ i.e. M = $\dfrac{n}{V}$

If we remember that n, is on the top line, the following may help us in calculations:

Since there is only one space for n, at the top of the triangle, M and V must be on the bottom

∴ M = $\dfrac{n}{V}$ or   n = M × V   or   V = $\dfrac{n}{M}$

Remember, V = Volume in *litres*

Worked Examples

1 How many moles are there in 100 cm³ of $\dfrac{M}{5}$ sodium hydroxide?

n = M × V = $\dfrac{1}{5}$ × $\dfrac{100}{1000}$ = 0.02

∴ No. of moles = 0.02 moles

2 What is the concentration of a solution of sulphuric acid which contains 0.1 moles in 50 cm³?

M = $\dfrac{n}{V}$ = $\dfrac{0.1 \times 1000}{50}$ = 2

∴ Molarity = 2M

3 What volume of a 2M sodium sulphate solution contains 0.5 moles sodium sulphate.

V = $\dfrac{n}{M}$ = $\dfrac{0.5}{2}$ = 0.25 litres

∴ Volume of solution = 0.25 1 = 250 cm³.

Formulae and the mole

Some Revision questions

1 Try the following problems.
   (a) 5.6 g iron was found to react with 1.6 g oxygen. Find the formula of the iron oxide.
   (b) 1.2 g magnesium was burned, and formed 2 g of magnesium oxide. What is the formula of the magnesium oxide?
   (c) 1.6 g of sulphur was burned in air. The weight of sulphur dioxide formed was 3.2 g.
   What is the formula of the sulphur dioxide? (Section 4.7.)

2 Calculate the percentage of nitrogen in each of the following compounds and hence, since nitrogen compounds are important as fertilisers, pick out the best fertiliser. (Section 4.5.)

   (i) Ammonium sulphate, (ii) ammonium nitrate, (iii) ammonia (NH_3).

3 What is (i) a mole, (ii) a molar solution? (Section 4.6.)
4 How many moles are
   (a) 112 g of iron;
   (b) 150 g of calcium carbonate;
   (c) 14.2 g of sodium sulphate;
   (d) 16 g of oxygen gas (O_2)? (Section 4.6.)
5 How many grams of each of the following substances would you weigh out in order to make up the following solutions:
   (a) 1 litre of 2M sodium chloride;
   (b) 1 litre of $\frac{M}{2}$ sodium sulphate;
   (c) 100 cm^3 (0.1 litre) of M silver nitrate;
   (d) 100 cm^3 of $\frac{M}{10}$ sodium hydroxide? (Section 4.8.)

6 How many moles are contained in:
   (a) 1 litre of 3M potassium hydroxide;
   (b) 100 cm^3 of M ammonium chloride;
   (c) 50 cm^3 of 0.5M sulphuric acid;
   (d) 100 cm^3 of 0.1M sodium chloride? (Section 4.8.)

7 What is the molarity of the following solutions:
   (a) 6 moles in 2 litres of solution;
   (b) 1 mole in 500 cm^3 of solution;
   (c) 0.1 moles in 100 cm^3 of solution;
   (d) 2 moles in 4 litres of solution? (Section 4.8.)

5

Equations

An equation is a means of summing up a chemical reaction.

5.1 Unbalanced equations

These can be used as a shorthand method of stating the change taking place in a reaction.
For example: $MgO + HCl \longrightarrow MgCl_2 + H_2O$

In other words, magnesium oxide plus hydrochloric acid forms magnesium chloride plus water.

This equation tells us *nothing* about quantities used or the conditions of the experiment. However, *this unbalanced equation is often all that is required.*

5.2 Balanced equations

These give us information about quantities used.

For example: $MgO + 2HCl \longrightarrow MgCl_2 + H_2O$
                1 mole 2 moles   1 mole   1 mole

This tells us that for a complete reaction, we need 2 moles of hydrochloric acid for every 1 mole of oxide. (See Fig. 18.)

5.3 Using equations in chemical calculations

If we write a balanced equation, we can use this equation to predict the quantities of a material required for an experiment, or to predict what quantities of the products would be obtained.

Examples

(a) What weight of carbon dioxide would you expect from adding

Equations

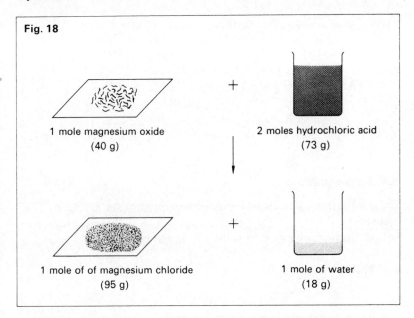

Fig. 18

excess hydrochloric acid to 5 g of calcium carbonate?
Balanced equation:

$CaCO_3 + 2HCl \longrightarrow CaCl_2 + H_2O + CO_2$
1 mole   1 mole

From the equation, we see that 1 mole of $CaCO_3$ gives 1 mole of CO_2.

That is, (40+12+48) g $CaCO_3$ gives (12+32) g CO_2
or   100 g $CaCO_3$ gives   44 g CO_2
Therefore,   5 g $CaCO_3$ gives $\dfrac{44 \times 5}{100}$ g CO_2

            = 2.2 g CO_2

(b) How many moles of hydrochloric acid would be required to react with 10 g magnesium oxide?
Balanced equation:

$MgO + 2HCl \longrightarrow MgCl_2 + H_2O$
1 mole   2 moles

From the equation, we see that:
    1 mole of MgO requires 2 moles of HCl
That is, (24+16) g of MgO requires 2 moles of HCl
or,   40 g of MgO requires 2 moles of HCl
    Therefore, 10 g MgO requires $\frac{1}{2}$ mole of HCl
                       = 0.5 moles of HCl

Your teacher may give you some problems to try for yourself.

5.4 State equations

State equations give us more information about a reaction. The letters (s), (l), (g), or (aq) are used to tell us if the materials are solids, liquids, gases, or in solution in water (aqueous).

e.g. $MgO(s) + 2HCl(aq) \longrightarrow MgCl_2(aq) + H_2O(l)$

5.5 Ionic equations (with more information)

We can rewrite the balanced state equation above showing the ions which are present.

$Mg^{2+}O^{2-}(s) + 2H^+Cl^-(aq) \longrightarrow Mg^{2+}(Cl^-)_2(aq) + H_2O(l)$

Since ionic compounds in solution are present as free ions, we could expand this equation even further, to give:

$Mg^{2+}O^{2-}(s) + 2H^+(aq) + 2Cl^-(aq) \longrightarrow Mg^{2+}(aq) + 2Cl^-(aq) + H_2O(l)$

Here we see that some species have remained unchanged throughout the reaction — in this case the Mg^{2+} ions and the Cl^- ions have remained as ions.
    These ions have not taken any part in the reaction, and are therefore called *spectator ions*.
The spectator ions can be omitted from the equation:

$O^{2-}(s) + 2H^+(aq) \longrightarrow H_2O(l)$

This equation tells us that 1 mole of oxide ions added to

Equations

2 moles of hydrogen ions (in *any* dilute acid), forms 1 mole of water.

Another example is that of dilute sulphuric acid and sodium hydroxide solution reacting to form sodium sulphate solution and water.

Unbalanced equation: $H_2SO_4 + NaOH \longrightarrow Na_2SO_4 + H_2O$

Balanced state equation:
$H_2SO_4(aq) + 2NaOH(aq) \longrightarrow Na_2SO_4(aq) + 2H_2O(l)$

Ionic equation:
$(H^+)_2SO_4^{2-}(aq) + 2Na^+OH^-(aq) \longrightarrow (Na^+)_2SO_4^{2-}(aq) + 2H_2O(l)$

In order to see more clearly which ions are the spectator ions, we can expand the equation to show free ions in solution.

i.e. $2H^+(aq) + SO_4^{2-}(aq) + 2Na^+(aq) + 2OH^-(aq)$
$\longrightarrow 2Na^+(aq) + SO_4^{2-}(aq) + 2H_2O(l)$

We see that the $Na^+(aq)$ and the $SO_4^{2-}(aq)$ ions are spectator ions, and if these are omitted the equation could now read:

$$2H^+(aq) + 2OH^-(aq) \longrightarrow 2H_2O(l)$$
$$\text{or } H^+(aq) + OH^-(aq) \longrightarrow H_2O(l)$$

That is, the reaction taking place is only between the hydrogen ions of the acid and the hydroxide ions of the alkali, forming water.

5.6 The formula of acids

The formula of hydrogen chloride has earlier been written as HCl, and it has been thought of as a covalent material. However, in 2.1 you learned that acids such as hydrochloric acid were electrolytes and therefore contained free ions.

When hydrogen chloride, HCl (g), dissolves in water hydrochloric acid, HCl(aq), is obtained and ions are produced, $H^+Cl^-(aq)$.

This is explained in chapter 7, but meantime we can consider all acids in solution in water as containing hydrogen ions, $H^+(aq)$.

Dilute hydrochloric acid, HCl(aq), consists of H^+Cl^-(aq).
Dilute sulphuric acid, H_2SO_4(aq), consists of $(H^+)_2SO_4^{2-}$(aq).
Dilute nitric acid, HNO_3(aq), consists of $H^+NO_3^-$(aq).

Some revision questions

You should try to answer each question and then check your answer by referring to the section indicated in brackets after the question.

1 When solid calcium carbonate is added to dilute hydrochloric acid, carbon dioxide is given off, and the calcium chloride and water are formed.
  (a) Write an unbalanced equation for the above reaction.
  (b) Write the balanced equation.
  (c) How many moles of calcium carbonate would you have to use to get 0.2 moles of carbon dioxide? (Assume that there is sufficient acid for the reaction).
  (d) What weight of calcium carbonate would this be?
  (e) What weight of calcium chloride would be obtained?
(Section 5.1 to 5.3. For the formulae refer to 4.1 and 4.2.)

2 When zinc is added to copper (II) sulphate solution, copper metal is formed and the zinc goes into solution as zinc sulphate.
  (a) Write the balanced state equation for this reaction.
  (b) Write the ionic equation.
  (c) Rewrite the ionic equation, this time omitting spectator ions. (Section 5.4 and 5.5.)

6

Activity and the electrochemical series

6.1 Revision of experiments with metals

(a) Burning metals in oxygen

Using the apparatus in Fig. 19, we can find an *order of activity* for the metals.

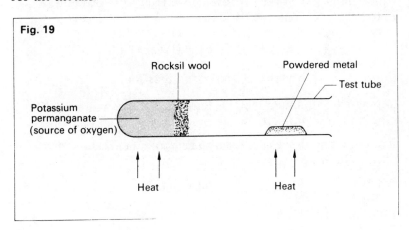

Fig. 19

Magnesium	: $2Mg + O_2 \rightarrow 2MgO$	↑	most reactive
Aluminium	: $4Al + 3O_2 \rightarrow 2Al_2O_3$		
Zinc	: $2Zn + O_2 \rightarrow 2ZnO$		
Iron	: $4Fe + 3O_2 \rightarrow 2Fe_2O_3$		
Lead	: $2Pb + O_2 \rightarrow 2PbO$		
Copper	: $2Cu + O_2 \rightarrow 2CuO$		least reactive

(b) Metals and water

By placing small pieces of metal in water, the following order of activity can be found (most reactive first);

Potassium : $2K + 2H_2O \rightarrow 2KOH + H_2$
Sodium   : $2Na + 2H_2O \rightarrow 2NaOH + H_2$
Lithium   : $2Li + 2H_2O \rightarrow 2LiOH + H_2$
Calcium   : $Ca + 2H_2O \rightarrow Ca(OH)_2 + H_2$

Hydrogen and an alkali are formed.
Other metals react slowly, if at all, with water.

(c) Metals and acid

(Note: potassium, sodium, lithium and calcium are *never* added to acid.) If other metals are reacted with dilute acids, the following order of activity is found (most reactive first):

Magnesium : $Mg(s) + 2HCl(aq) \rightarrow H_2(g) + MgCl_2(aq)$.
Aluminium : $2Al(s) + 6HCl(aq) \rightarrow 3H_2(g) + 2AlCl_3(aq)$.
Zinc   : $Zn(s) + 2HCl(aq) \rightarrow H_2(g) + ZnCl_2(aq)$.
Iron   : $Fe(s) + 2HCl(aq) \rightarrow H_2(g) + FeCl_2(aq)$.
Tin   : $Sn(s) + 2HCl(aq) \rightarrow H_2(g) + SnCl_2(aq)$.

Copper   : $Cu(s) + H^+(aq) \rightarrow$ no reaction.

In practice aluminium metal is fairly unreactive with acid. This is because it is protected by an oxide layer — discussed later.

By combining the results from these experiments, an *activity series* is built up.

Potassium	K	↑	most active
Sodium	Na		
Lithium	Li		
Calcium	Ca		
Magnesium	Mg		
Aluminium	Al		
Zinc	Zn		
Iron	Fe		
Tin	Sn		
Lead	Pb		
Copper	Cu		least active

6.2 Displacement of one metal by another

Active metals can *displace* less active metals from their salts. For

Activity and the electrochemical series

example, magnesium will react with copper sulphate solution to produce magnesium sulphate and copper. In other words, the magnesium has displaced the copper from the copper sulphate solution:

$$Mg(s) + CuSO_4(aq) \rightarrow MgSO_4(aq) + Cu(s)$$

or, if we omit spectator ions, the equation will read:

$$Mg(s) + Cu^{2+}(aq) \rightarrow Mg^{2+}(aq) + Cu(s)$$

Similarly: $Fe(s) + Cu^{2+}(aq) \rightarrow Fe^{2+}(aq) + Cu(s)$
But: $Pb(s) + Mg^{2+}(aq) \nrightarrow$ no reaction, since lead is less active than magnesium.

Similarly: $Mg(s) + Mg^{2+}(aq) \nrightarrow$
$Cu(s) + Cu^{2+}(aq) \nrightarrow$ } no reaction

Displacement, and reacting masses

From the *balanced* equation

$$Mg(s) + Cu^{2+}(aq) \rightarrow Mg^{2+}(aq) + Cu(s)$$

We see that 1 mole of magnesium should displace 1 mole of copper from Cu^{2+} solution.
That is, 24 g Mg should displace 64 g Cu from Cu^{2+} ions in solution.
That is, 1 g Mg should displace $\frac{64}{24}$ = 2.67 g Cu.
This result may be verified by experiment.

6.3 Ion-electron equations, and electron transfer

The equation, $Mg(s) + Cu^{2+}(aq) \rightarrow Mg^{2+}(aq) + Cu(s)$ can be split up into 2 equations, describing the changes taking place. (These ion-electron equations can be obtained from the data tables on page 144.)

$Mg(s) \rightarrow Mg^{2+}(aq) + 2e$
$Cu^{2+}(aq) + 2e \rightarrow Cu(s)$

Since the magnesium atoms have formed magnesium ions, they must have lost 2 electrons per atom, and the copper ions must have picked up 2 electrons in forming copper atoms.

Electrons transfer must have taken place from the magnesium atoms to the copper ions.

This may be demonstrated by an experiment similar to that in Fig. 20.

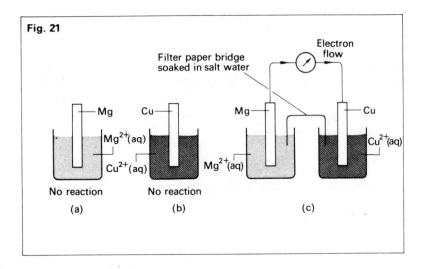

Fig. 21

In Fig. 20(c), magnesium forms magnesium ions in solution, and fresh copper is deposited on the copper electrode.

6.4 Reduction of oxides

	Oxide of	
Increasing ease of reduction	Potassium Sodium Calcium Magnesium Aluminium Zinc	Not readily reduced to the metal.
	Iron Tin Lead Copper	Reduced to the metal by the action of hydrogen or carbon on hot oxide.
	Mercury Silver	Decomposed by heat alone.

Since the most active metal is the one which forms ions most readily (e.g. K → K⁺ + e), then the ions formed will be least likely to pick up electrons to reform the metal. (That is, $K^+ + e \rightarrow K$ will be difficult to achieve.) Therefore, the lower a metal is in the activity series, the easier it is to obtain from its oxide.

6.5 Ease of forming ions in solution

All metals tend to form ions in solution. When zinc metal is placed in water, it tends to form zinc ions:
$Zn(s) \rightarrow Zn^{2+}(aq) + 2e$

The electrons are left on the zinc plate, giving it a negative charge (Fig. 21).

Similarly, with copper, some copper ions will form and the metal will develop a negative charge due to electrons being left behind (Fig. 22).

However, all metals do not form ions equally readily. In the above case, the zinc forms ions more readily than the copper. As a result, there will be a bigger 'build up' of electrons (that is, a higher electrical energy or *potential*) on the zinc plate.

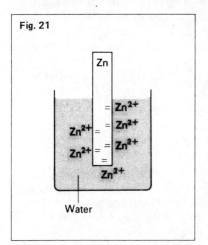

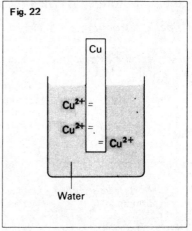

If the two strips are connected by a wire, the potential difference (voltage) between the plates forces electrons to flow from the zinc to the copper. A voltmeter can measure this potential difference (Fig. 23).

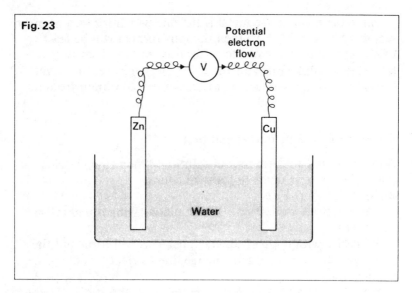

Fig. 23

The potential difference between different metal pairs can be measured. For our purposes, copper metal can be a convenient standard to use (although it is not used by chemists as the standard) and if the metals are arranged in order of voltage, the series formed is called the Electrochemical Series.

The Electrochemical Series.

↑ Metals *most readily* forming M^+ (aq) ions	Lithium: Li(s) ⇌ Li^+(aq) + e Potassium: K(s) ⇌ K^+(aq) + e Calcium: Ca(s) ⇌ Ca^{2+}(aq) + 2e Sodium: Na Magnesium; Mg Aluminium; Al Zinc, Zn Iron, Fe Tin, Sn Lead, Pb	Metal ions *least readily* form the metal
Metals *least readily* forming M^+ (aq)	Copper: Cu ⇌ Cu^{2+}(aq) + 2e Mercury: Hg ⇌ Hg^{2+}(aq) + 2e	Metal ions *most readily* form the metal

Note: the series is very similar to the activity series, *but* lithium is at the top in this series, and the positions of sodium and calcium are reversed.

The Electrochemical Series compares the *energy* involved in converting metal atoms into metal ions in solution.

The Reactivity Series compares the *rates* of reactions involving metals. In other words, lithium may take longer to react than potassium, but the total energy released on forming ions in solution is greater.

6.6 Oxidation and reduction — redox reactions

Oxidation used to be thought of as gain of oxygen, for example:

$2Mg + O_2 \rightarrow 2MgO$

The magnesium atoms have formed magnesium ions, and lost electrons in the process:

$Mg \rightarrow Mg^{2+} + 2e$

When magnesium burns in chlorine, magnesium chloride is formed:

$Mg + Cl_2 \rightarrow MgCl_2$

Here again the magnesium atoms have lost electrons and formed ions:

$Mg \rightarrow Mg^{2+} + 2e$

Since the same change has taken place in both these reactions (the magnesium has lost electrons and formed ions), the meaning of oxidation has been extended to cover all such changes.
Oxidation is the process of electron loss.
(Remember: Loss of Electrons is Oxidation — **LEO**.)
The reverse process, electron gain, is called reduction.
Reduction is the process of electron gain.
(Remember: Reduction is Electron Gain — **REG**.)

Since both electron loss (oxidation) and electron gain (reduction) must take place in the same reaction, an oxidation-reduction reaction is usually called a *redox* reaction.

We have already discussed the electron transfer from magnesium atoms to copper ions; the reactions taking place being:

$Mg(s) \rightarrow Mg^{2+}(aq) + 2e$ Oxidation
$Cu^{2+}(aq) + 2e \rightarrow Cu(s)$ Reduction

Cells can be set up to demonstrate electron transfer in other redox reactions. For example (Fig. 24):

The Fe^{3+} ions in iron (III) chloride solution react with I^- ions in potassium iodide solution. The Fe^{3+} ions are reduced to Fe^{2+} ions, and the I^- ions are oxidised to I_2 (iodine):

$Fe^{3+} + e \rightarrow Fe^{2+}$ Reduction
$2I^- \rightarrow I_2 + 2e$ Oxidation

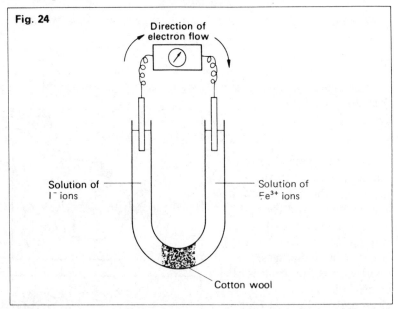

Fig. 24

The following tests can prove that Fe^{3+} ions are being converted to Fe^{2+} ions:

Fe^{3+}(aq) present : potassium thiocyanate turns red.
Fe^{2+}(aq) present : potassium hexacyanoferrate (III) turns blue.

Proof that iodine is formed can be shown by adding a little starch solution. A deep blue colour is obtained.

Activity and the electrochemical series

Redox reactions in electrolysis

Let us look at the electrode reactions during electrolysis of a number of compounds.

	Negative Electrode	Positive Electrode
1 Copper (II) chloride solution, $CuCl_2$	$Cu^{2+} + 2e \rightarrow Cu$	$2Cl^- \rightarrow Cl_2 + 2e$
2 Molten lead bromide, $PbBr_2$	$Pb^{2+} + 2e \rightarrow Pb$	$2Br^- \rightarrow Br_2 + 2e$
3 Extraction of aluminium from molten bauxite, Al_2O_3	$Al^{3+} + 3e \rightarrow Al$	$2O^{2-} \rightarrow O_2 + 4e$
	Electron gain, that is **reduction** at the negative electrode	Electron loss, that is **oxidation** at the positive electrode

6.7 Corrosion

When metals react with their natural surroundings (air or water) they form ions, and in so doing lose electrons.

This process is often called corrosion.

Corrosion of iron

Experiments show that before iron will rust, both water and air must be present.

Since iron is the major metal used in the world, we must find efficient methods of protecting it.

Two common methods used are: (a) galvanising (zinc plating);
(b) tin plating.

Efficiency of galvanising and tin plating

Both galvanising and tin plating are effective in stopping corrosion of iron so long as the layers are complete. However, when the surfaces are scratched there is a considerable difference:

Galvanised iron — there is still little or no corrosion of the iron.
Tin plated iron — the iron now corrodes very quickly.

This suggests that the process of rusting (iron forming ions) may either be assisted or prevented, depending on what is connected to the iron.

The electrochemical nature of corrosion

When iron is connected to another metal, electrons will either flow to the iron, or from it, depending on the position of the other metal in the electrochemical series. Ferroxyl indicator (a mixture of phenolphthalein and potassium hexacyanoferrate (III)) is useful in investigations since in the presence of iron (II) ions it turns blue, and in the presence of hydroxide ions it turns pink.

Experiments similar to those in Fig. 25 can be carried out to find out more about the corrosion process.

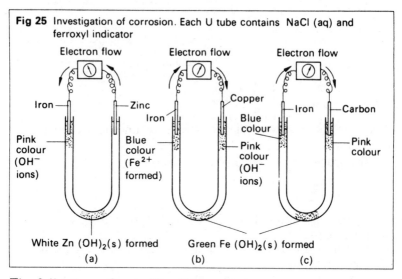

Fig 25 Investigation of corrosion. Each U tube contains NaCl (aq) and ferroxyl indicator

The following refers to Fig. 25.
(a) Iron and zinc
   The zinc forms ions more readily than the iron, and so goes into solution as zinc ions.

   $$Zn(s) \longrightarrow Zn^{2+}(aq) + 2e$$

   The electrons are pushed on to the iron, which does not corrode. Instead, hydroxide ions form at the iron.
(b) Iron and copper
   Iron forms ions more readily than copper.

   $$Fe(s) \longrightarrow Fe^{2+}(aq) + 2e$$

   Electrons flow from the iron to the copper, and the hydroxide ions form at the copper.
(c) Iron and carbon
   Since carbon is used in the reduction of iron ore in the blast

Activity and the electrochemical series

furnace, there is always carbon as impurity in iron, and it is important to study its effect. Once again the iron ionises.

$$Fe(s) \longrightarrow Fe^{2+}(aq) + 2e$$

The electrons flow from the iron to the carbon, and hydroxide ions form at the carbon.

We can now conclude from the experiment that if electrons are pushed on to the iron it will not form ions, and if the electrons are encouraged to leave the iron, it will form ions (corrode) quickly.

The normal rusting of iron

Due to carbon impurity, corrosion of iron is speeded up by the formation of an iron/carbon cell as in Fig. 25(c).

$$Fe \longrightarrow Fe^{2+} + 2e$$

The electrons flow to the carbon and hydroxide ions form at the carbon (Fig. 26).

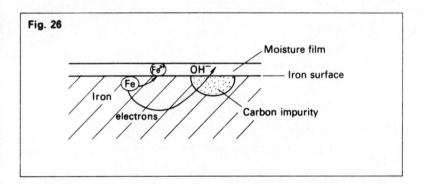

Fig. 26

How the hydroxide ions form when iron rusts

You may wonder how the hydroxide ions are formed during the corrosion process. Experiments indicate that both air and water are required for the corrosion of iron. Since the corrosion of iron is the process of iron forming ions and losing electrons,

$$Fe \longrightarrow Fe^{2+} + 2e$$

then the formation of hydroxide ions must involve the gain of electrons by the water and oxygen of the air.

$$2H_2O + O_2 + 4e \longrightarrow 4OH^-$$

This equation can be found written in reverse in the data tables on page 145.

6.8 Protecting iron

Any method where electrons are pushed on to the iron should give good protection from corrosion. Methods used are:—

1 Cathodic protection (Fig. 27)

The iron is connected to the negative terminal of a D.C. power supply. Electrons are pushed on to the iron, preventing it from forming ions.

This can be used to protect steel piers, etc.

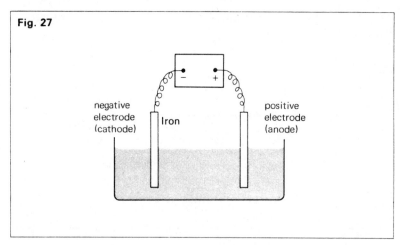

Fig. 27

2 Sacrificial protection, using magnesium

Magnesium scrap is connected to the iron, and since it forms ions more readily, electrons are pushed on to the iron, stopping it ionising.

Activity and the electrochemical series

Oil pipe-lines, or steel constructions too expensive to replace, can be protected in this manner.

3 Galvanising (zinc plating)

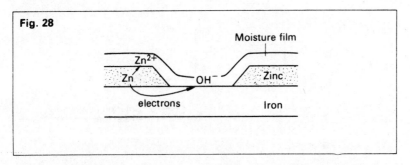

Fig. 28

The zinc protects the iron from air and water. When the surface is scratched, and the iron is exposed, the zinc still protects the iron by sacrificial protection (as in Fig. 28).

4 Tin plating (as in 'tin cans')

The tin protects the iron physically. However, when there is a break in the tin, the iron forms ions (corrodes) more rapidly, since tin is less active than iron, and electrons flow from the iron to the tin.

However, tin plating is still used for canning foods because it is much less active than zinc, and is less likely to react with the food if it comes into contact with it.

5 Electroplating

Copper, nickel or chromium are plated on to the iron by making the iron the cathode (negative terminal) of a cell containing the metal salt as electrolyte (Fig. 29).

6 Painting or Greasing

Effect of salt on corrosion rate

Since the process of corrosion involves the setting up of

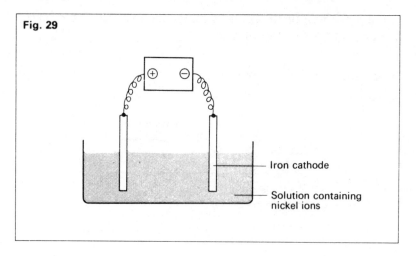

electrolytic cells, an ionic solution (for example, sodium chloride), which is a better conductor than water, will speed up the corrosion rate.

6.9 Aluminium

Aluminium is more reactive than iron, but it does not corrode readily. This is because a thin oxide layer forms on the surface, preventing attack from air and water, and so protecting the aluminium from further corrosion.

If aluminium is dipped in mercury (II) chloride solution, some of the mercury ions are displaced, and the mercury atoms formed go on to the aluminium. This does not stop the aluminium oxidising in air, but it does stop the oxide layer sticking to the aluminium surface. When left exposed to air, the aluminium now corrodes very quickly, and flakes of aluminium oxide form.

Anodising aluminium

By this process, the oxide layer on the aluminium is thickened, protecting the aluminium even further from corrosion.

The aluminium is made the positive electrode *(anode)* of a cell containing dilute sulphuric acid. When a current is passed through, oxygen is released at the aluminium anode, and thickens the oxide layer on the surface.

Activity and the electrochemical series

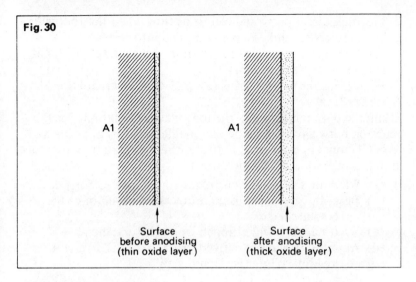

Fig. 30

Surface before anodising (thin oxide layer)

Surface after anodising (thick oxide layer)

Uses of aluminium

1 The anodised aluminium can be used out of doors without fear of corrosion.
2 Anodised aluminium foil is used in cooking.
3 The thick oxide layer can absorb dyes, and is used for making coloured aluminium pots and pans, and coloured wrappings (e.g. for chocolate biscuits).
The dyeing of aluminium can be demonstrated in the laboratory.

Some revision questions

You should try to answer each question, and then check your answer by referring to the section indicated in brackets after the question.
   The electrochemical series is given in the Reduction Potential Tables at the end of this book. You should use them to answer much of the following.
1 What would happen in the following cases? Give the ion-electron equations for any reaction which takes place.
   (a) Magnesium added to a solution of copper (II) ions.
   (b) Zinc added to a solution of lead ions.
   (c) Iron added to a solution of zinc ions. (Sections 6.2 and 6.3.)

2 In which direction would electrons flow when the following pairs of metals are linked up and dipped into water?
(a) Zinc and copper; (b) magnesium and lead; (c) zinc and magnesium. (Section 6.5.)
3 What is meant by the terms (a) oxidation, (b) reduction, (c) redox?
Using the data tables, write the ion electron equations for the reaction between (i) chlorine and iodide ions, (ii) bromine and iron (II) ions. In each case, indicate where oxidation and reduction has taken place. (Section 6.6.)
4 (a) What are the conditions required for iron to corrode?
(b) Describe how galvanising protects iron even once the surface is scratched.
(c) What happens when tinned iron gets scratched?
(d) Aluminium is more active than iron, but it does not corrode readily. Why? (Section 6.7 to 6.9.)

7

Acids and bases

7.1 Acidic and basic oxides

When *non-metals* burn in oxygen, they often form *acidic oxides*. An acidic oxide forms an acidic solution if it dissolves in water (pH paper turns red).
Metals burn to form *basic oxides* (*bases*)
If these basic oxides (bases) dissolve in water, they form *alkaline* solutions (pH paper turns blue).
Many bases, however, do not dissolve in water and do not therefore form alkalis in water (for example, iron oxide and copper oxide).

7.2 Acids and the hydrogen ion

When acid solutions are electrolysed, hydrogen gas is always evolved at the cathode (negative electrode).
Therefore, acid solutions must contain the hydrogen ion, H^+(aq).

What is the hydrogen ion?

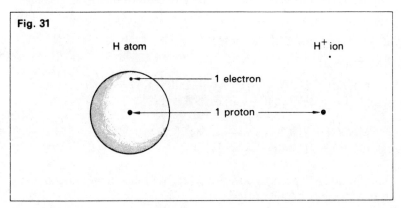

Fig. 31

The H^+ ion would simply be a *proton*, with no protective outer electron energy levels (Fig. 31).

The unprotected positive charge is very powerful and is automatically attracted to polar water molecules, becoming hydrated (surrounded by water molecules).

$$H^+ + H_2O \longrightarrow H^+(aq)$$

This happens in all acid solutions.

An acid is a hydrogen containing compound, which in solution in water forms hydrated hydrogen ions (H^+ (aq)).

Characteristics of Acids

1 Acids and metals

Dilute acids react with metals from magnesium to tin in the activity series (for safety metals above magnesium are never used with acids).

Hydrogen gas is given off. Examples are:

$$Mg(s) + H_2SO_4(aq) \longrightarrow MgSO_4(aq) + H_2(g)$$
i.e. $Mg(s) + 2H^+(aq) \longrightarrow Mg^{2+}(aq) + H_2(g)$

$$Fe(s) + 2HCl(aq) \longrightarrow FeCl_2(aq) + H_2(s)$$
i.e. $Fe(s) + 2H^+(aq) \longrightarrow Fe^{2+}(aq) + H_2(g)$

The hydrogen ions of the acid have been reduced to hydrogen.

2 Acids + metal oxides (bases)

Examples are:

$$CuO(s) + 2HNO_3(aq) \longrightarrow Cu(NO_3)_2 + H_2O$$
i.e. $O^{2-}(s) + 2H^+(aq) \longrightarrow H_2O(l)$

$$MgO(s) + 2HCl(aq) \longrightarrow MgCl_2(aq) + H_2O$$
i.e. $O^{2-}(s) + 2H^+(aq) \longrightarrow H_2O(l)$

The metal ions go into solution. The acid has been neutralized and water has been formed.

Acids and bases

3 Acid + carbonates

Carbon dioxide gas is released, and water is formed. This happens when the carbonate is present as a solid, or in solution. For example:

$$CuCO_3(s) + H_2SO_4(aq) \longrightarrow CuSO_4(aq) + CO_2 + H_2O$$
$$CO_3^{2-}(s) + 2H^+(aq) \longrightarrow CO_2(g) + H_2O(b)$$

Once again, the acid has been neutralized, and water has been formed.

4 Acid + alkali

In each case, water is formed. For example:

$$HCl(aq) + NaOH(aq) \longrightarrow NaCl(aq) + H_2O$$
$$H^+(aq) + OH^-(aq) \longrightarrow H_2O(l)$$

This is another example of neutralization.

7.3 Bases and the hydroxide ion, OH⁻

A *base* is usually a metallic oxide or hydroxide. Examples are copper oxide (CuO), sodium hydroxide (NaOH) and iron (II) hydroxide ($Fe(OH)_2$).

An *alkali* is the name given to the aqueous solution of a base. All alkalis contain the hydroxide ($OH^-(aq)$) ion.

e.g. $Ca^{2+}O^{2-}(s) + H_2O \longrightarrow Ca^{2+}(aq) + 2OH^-(aq)$
$Na^+OH^-(s) + H_2O \longrightarrow Na^+(aq) + OH^-(aq)$
Base   Alkali

Insoluble bases cannot form alkalis. For example, copper oxide and iron (II) hydroxide are both insoluble in water, so no free hydroxide ions can be formed.

7.4 The pH scale

The pH scale is used to define the concentration of hydrogen ions in solution. pH paper, or Universal indicator solution (pH indicator solution) is used to find the pH value of a solution.

0	1	2	3	4	5	6	7	8	9	10	11	12	13	14
increasing H^+(aq) concentration ←							neutral	increasing OH^- concentration →						

7.5 Strong and weak acids

When the conductivities of hydrochloric acid and ethanoic acid of equal concentrations are compared, the hydrochloric acid shows a much higher conductivity.

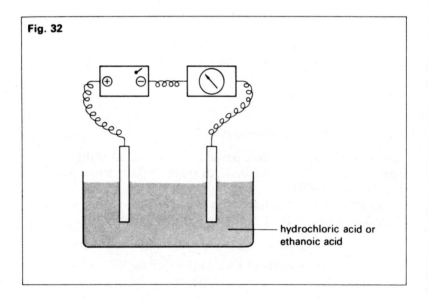

Fig. 32

hydrochloric acid or ethanoic acid

This indicates that ethanoic acid has fewer ions in solution.

Also, the pH of $\frac{M}{10}$ hydrochloric acid is found to be 1, whereas that of $\frac{M}{10}$ ethanoic acid is found to be 3, again indicating that ethanoic acid has fewer hydrogen ions in solution.

Hydrochloric acid is a *strong acid*. When dissolved in water, it is almost completely dissociated into ions.

Acids and bases

$HCl(g) + H_2O \longrightarrow H^+(aq) + Cl^-(aq)$
hydrogen   hydrochloric acid
chloride

Nitric and sulphuric acids are examples of other strong acids.

$HNO_3(l) + H_2O \longrightarrow H^+(aq) + NO_3^-(aq)$
$H_2SO_4(l) + H_2O \longrightarrow 2H^+(aq) + SO_4^{2-}(aq)$

Ethanoic acid is a *weak acid*. When dissolved in water, it is only dissociated to a very small extent. That is, it exists in solution mainly as undissociated molecules.

$CH_3.COOH + H_2O \rightleftharpoons CH_3COO^-(aq) + H^+(aq)$
mostly in this state   only a few ions present

7.6 Acid strength and concentration

Acid strength is the degree to which an acid will dissociate into ions in aqueous solution.
Concentration is the amount of acid dissolved in a litre of the solution. (The unit of concentration is molarity.)
   Try not to confuse the meaning of these terms.

7.7 Position of bases on the pH scale

7
8
9
10
11 .... lime water (calcium hydroxide)
12. .... ammonia
13
14 .... sodium hydroxide, potassium hydroxide.

Sodium hydroxide is a *strong base* (or strong alkali). That is, it is almost completely dissociated in solution (it is almost completely split up into free sodium ions and free hydroxide ions.)

Ammonia solution is a weak base — it only forms a small number of ions in solution.

7.8 The ionisation of water

We originally looked on water as being a polar covalent compound:

However, experiments show that pure water does conduct electricity slightly. This indicates the existence of a very small number of ions. In addition,

(a) When an acid solution is diluted, the pH eventually reaches 7.

Since the pH of pure water also equals 7, pure water must contain a very small number of hydrogen ions.

(b) When an alkali solution is diluted, the pH eventually reaches 7.

Pure water must therefore contain a very small number of hydroxide ions.

In other words, pure water must contain a very small number of hydrogen and hydroxide ions.

Water molecules must dissociate *to a very slight extent* into hydrogen and hydroxide ions:

$$H_2O(l) \rightarrow H^+(aq) + OH^-(aq)$$

most water is in this form.   only a very small number of ions present.

Acids and bases

Therefore water (pH = 7) must have equal numbers of hydrogen and hydroxide ions.

We also know that H^+ ions of an acid and OH^- ions of an alkali will neutralise each other to form water.

$$H^+(aq) + OH^-(aq) \rightarrow H_2O$$

The reaction is therefore *reversible*

$$H_2O(l) \rightleftharpoons H^+(aq) + OH^-(aq)$$

and in water an *equilibrium* is set up in which a few water molecules are always dissociating into ions, while the ions recombine to form water. At any one instant, the proportion of water molecules to hydrogen and hydroxide ions always remains constant.

What happens when hydrogen ions or hydroxide ions are removed from water?

When hydrogen ions are removed from water, the balance is upset, and an excess of hydroxide ions is produced (that is, the solution becomes alkaline). Similarly, when hydroxide ions are removed, an excess of hydrogen ions is produced (that is the solution becomes acidic).

Some revision questions

You should try to answer each question, and then check your answer by referring to the section indicated in brackets after the question.

1 What are acidic and basic oxides? (Section 7.1.)
2 List the main properties of an acid, and give examples of each. (Section 7.2.)
3 What is the connection between a base and an alkali? Why can all alkalis be called bases, but all bases cannot be called alkalis? (Section 7.3.)
4 (a) What is meant by the pH scale?
  (b) State the range of pH numbers which would indicate (i) an acid, (ii) an alkali, (iii) a neutral solution.
  (c) What is the difference between a *strong* and a *weak* acid? How can a *weak* acid be *concentrated*? (Section 7.4 and 7.5.)

5 Water conducts electricity to a very slight extent. Why? What is meant by saying that water molecules dissociate, and an equilibrium is set up? (Section 7.8.)

8

Neutralization and salt formation

8.1 What is a salt?

Salts all get their names from their parent acids.
For example, sulphuric acid (H_2SO_4) forms sulphates
hydrochloric acid (HCl) forms chlorides
nitric acid (HNO_3) forms nitrates

A salt also contains a positive ion — a positive metal ion or the ammonium, NH_4^+, ion.

We may, therefore, define a salt thus:
A salt is the substance formed when the hydrogen ions of an acid are replaced by metal ions (or the ammonium ion).

8.2 Solubility of ionic compounds

To decide the correct method of preparing a salt, we must know whether various compounds are soluble or insoluble in water.

To help us, we can use the data tables 'Guide to Solubility in Water' on page 146.

The table uses codes for how soluble a material is. However, for our purposes at the moment, the salt can be treated as "insoluble" if its solubility is less than 10 gram per litre.

i.e. Code   'a'   soluble

Codes   'b', 'c' or 'i'   insoluble.

Note From the table you can see that the following are **always** soluble:

(i) Compounds of Metals in column 1 of the Periodic Table (e.g. sodium and potassium).
(ii) Ammonium compounds.
(iii) Nitrates.

8.3 Preparation of salts

(a) Soluble salts

1 *Acid + active metal* (from magnesium to tin in the Activity Series). Take the preparation of magnesium sulphate as an example.

$$Mg(s) + H_2SO_4(aq) \longrightarrow MgSO_4(aq) + H_2(g)$$

Crystals of magnesium sulphate are obtained by concentrating the solution (boiling off most of the water), and leaving to crystallise.

2 *Acid + basic oxide.* Take the preparation of copper (II) sulphate as an example.

$$CuO(s) + H_2SO_4(aq) \rightarrow CuSO_4(aq) + H_2O$$

Blue copper sulphate crystals are obtained by filtering off excess copper oxide, concentrating the solution, and leaving to crystallise.

3 *Acid + alkali.* Take the preparation of sodium chloride as an example.

$$NaOH(aq) + HCl(aq) \rightarrow NaCl(aq) + H_2O$$

Since all materials are in solution, excess acid or alkali must not be added. The 'end point' must therefore be found by using an indicator

4 *Acid + carbonate.* An alternative preparation of copper sulphate provides an example.

$$H_2SO_4(aq) + CuCO_3(s) \rightarrow CuSO_4(aq) + H_2O + CO_2(g)$$

Carbon dioxide gas is evolved, and the copper sulphate crystals are obtained from the solution as before, by filtering off the excess copper carbonate, concentrating the solution, and leaving to crystallise.

Neutralization and salt formation

(b) Insoluble salts — prepared by precipitation

If a salt is insoluble it is prepared by adding a solution of the positive metal ions to a solution of the negative ions.

Take the preparation of barium sulphate as an example.

$$Ba^{2+}(aq) + SO_4^{2-}(aq) \longrightarrow Ba^{2+}SO_4^{2-}(s)$$

The barium and sulphate ions may come from a variety of compounds which contain them.

e.g. $BaCl_2(aq) + Na_2SO_4(aq) \longrightarrow BaSO_4(s) + 2NaCl(aq)$
     soluble   soluble   insoluble
     barium
     chloride

or $Ba(OH)_2(aq) + H_2SO_4(aq) \longrightarrow BaSO_4(s) + 2H_2O$
    soluble

Whenever a barium ion and a sulphate ion come together, water is unable to separate them again, and an insoluble precipitate is formed.

Only in one of the neutralization reactions indicated above is it obvious that the hydrogen ions are being removed from the acid, since hydrogen gas is liberated. In the other cases, we have only *assumed* that water is formed in using up the hydrogen ions. We must try to find some experimental evidence to prove this.

For a summary on deciding the method to use in salt preparations, see Appendix page 138.

8.4 Mobility of ions — another approach to neutralization

Earlier experiments on conductivity (2.1, 2.2) may have shown that not all electrolytes have equal conductivities.

Further experiments can suggest possible explanations.

(a) Concentration effect

In general we find that the *higher the concentration of the ions, the higher the conductivity.*

(b) Mobility (speed) of ions

To investigate effects other than concentration, we must use different solutions of equal concentration.

Results obtained from experiments on some of these solutions would be:
  (i) hydrochloric acid   (H^+ + Cl^-):   best conductor
  (ii) sodium hydroxide   (Na^+ + OH^-)
  (iii) sodium chloride   (Na^+ + Cl^-):   poorest conductor.

Comparing the results of (i) and (iii), the chloride ion (Cl^-) is common to both, and any differences must be due to the hydrogen (H^+) or sodium (Na^+) ions.

Therefore, the hydrogen ion (H^+) is faster than the sodium ion (Na^+).

In the same way, comparing (ii) and (iii), the Na^+ ion is common, therefore the hydroxide ion (OH^-) is faster than the chloride ion (Cl^-).

(c) Comparing the speeds of hydrogen ions and hydroxide ions

If an experiment is set up in which an indicator shows the movement of hydrogen ions towards the negative electrode, and of the hydroxide ions towards the positive electrode, then the hydrogen ion is seen to be about twice as fast as the hydroxide ion.

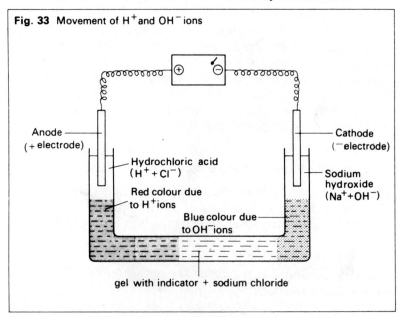

Fig. 33 Movement of H^+ and OH^- ions

To sum up: hydrogen ions are faster than hydroxide ions which are faster than all the others.

Therefore, if hydrogen ions or hydroxide ions are lost from a

Neutralization and salt formation 67

solution (perhaps being replaced by slower moving ions) then the conductivity of the solution should decrease.
The experiments on salt preparations could now be repeated, using conductivity measurements.

8.5 Magnesium added to dilute hydrochloric acid

The graph in Fig. 34 is obtained. The reaction is:

$$2H^+Cl^-(aq) + Mg(s) \longrightarrow Mg^{2+}(Cl^-)_2(aq) + H_2(g)$$

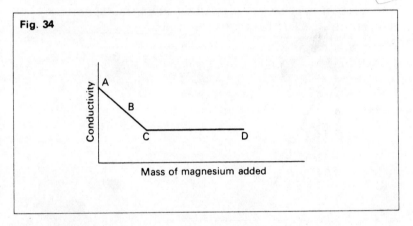

Fig. 34

Explanation of graph

	Ions in solution	
At A No magnesium has been added, so there are only H^+(aq) and Cl^-(aq) ions present. H^+ ions are fast. Therefore, the conductivity is high.	H^+ H^+ H^+ H^+	Cl^- Cl^- Cl^- Cl^-
At B Some magnesium has been added. This reacts, and replaces some of the H^+ ions with Mg^{2+} ions. The H^+(aq) ions form hydrogen gas. Therefore, the current drops.	Mg^{2+} H^+ H^+	Cl^- Cl^- Cl^- Cl^-
At C Sufficient magnesium has now been added to neutralize the acid, and all the fast moving H^+ ions have been replaced by slower moving Mg^{2+} ions. Therefore, the current reaches its lowest point.	Mg^{2+} Mg^{2+}	Cl^- Cl^- Cl^- Cl^-
From C to D The conductivity remains steady, since there are no more H^+ ions to be replaced, and the magnesium has only a very slow reaction with water.	no change Mg^{2+} Mg^{2+}	Cl^- Cl^- Cl^- Cl^-

Since $Cl^-(aq)$ ions are unchanged throughout, the reaction can be written:

$$Mg(s) + 2H^+(aq) \rightarrow Mg^{2+}(aq) + H_2(g)$$

8.6 Acid + alkali (sodium hydroxide solution added to hydrochloric acid)

The graph in Fig. 35 is obtained. The reaction is:

$$Na^+OH^-(aq) + H^+Cl^-(aq) \rightarrow Na^+Cl^-(aq) + H_2O$$

The reaction can also be followed by adding methyl orange indicator to the acid before the reaction.

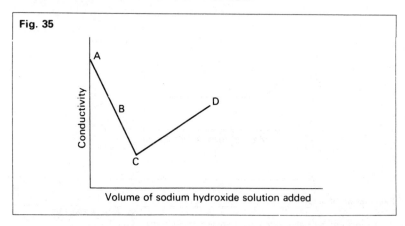

Fig. 35

Volume of sodium hydroxide solution added

Explanation of Graph

	Ions in solution
At A H^+ and Cl^- ions present.	H^+ H^+ Cl^- Cl^-
At B Half the NaOH required has been added. $H^+(aq)$ replaced by $Na^+(aq)$, due to $H^+(aq) + OH^-(aq) \rightarrow H_2O$. Therefore, the conductivity drops.	Na^+ Cl^- H^+ Cl^-
At C The **end point** (that is, reaction is complete). All the H^+ ions have now been replaced, so the conductivity is at its lowest. The indicator changes colour at this point, showing that neutralization is complete.	Na^+ Cl^- Na^+ Cl^-
C to D Excess NaOH being added. No H^+ left. OH^- ions now present (fast moving), so the conductivity rises.	Na^+ Cl^- Na^+ Cl^- Na^+ OH^-

Neutralization and salt formation

Note: Since the $Na^+(aq)$ and $Cl^-(aq)$ ions are spectator ions, the reaction may be written

$$H^+(aq) + OH^-(aq) \longrightarrow H_2O(l)$$

8.7 Acid + carbonate (solid copper (II) carbonate added to sulphuric acid solution)

The graph in Fig. 36 is obtained. The reaction is

$$(H^+)_2SO_4^{2-}(aq) + Cu^{2+}CO_3^{2-}(s) \rightarrow Cu^{2+}SO_4^{2-}(aq) + H_2O + CO_2(g)$$

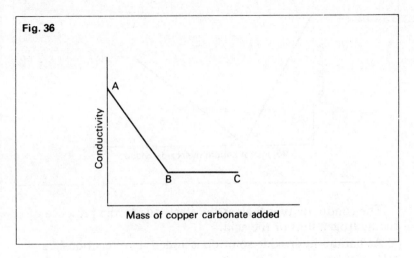

Fig. 36

Explanation of graph

	Ions in solution	
At A High conductivity, due to H^+ ions	H^+ H^+	SO_4^{2-}
At B All the H^+ ions have been replaced, and the only ions in solution are Cu^{2+} and SO_4^{2-}	Cu^{2+}	SO_4^{2-}
B to C As excess copper carbonate is added, there is no further change in conductivity since it is insoluble in water.	Cu^{2+}	SO_4^{2-}

If we omit spectator ions, the reaction is:

$$2H^+(aq) + CO_3^{2-}(s) \rightarrow H_2O + CO_2(g)$$

8.8 Precipitation (barium hydroxide solution added to dilute sulphuric acid)

The graph in Fig. 37 is obtained. The reaction is:

$(H^+)_2 SO_4{}^{2-}(aq) + Ba^{2+}(OH^-)_2(aq) \rightarrow Ba^{2+}SO_4^{2-}(s) + 2H_2O$

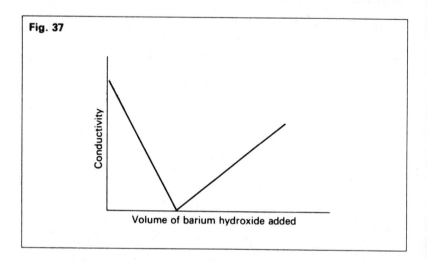

Fig. 37

The conductivity is high at the start, due to the presence of the hydrogen ions of the acid.

As barium hydroxide solution is added, two reactions are taking place:

$H^+(aq) + OH^-(aq) \rightarrow H_2O$ (H^+ ions removed from solution)
    (being added)

$SO_4{}^{2-}(aq) + Ba^{2+}(aq) \rightarrow Ba^{2+}SO_4{}^{2-}(s)$
    (being added)
                    insoluble, so
                      $SO_4{}^{2-}$ ions are being removed
                      from solution.

Therefore, at the *end point*, there are no ions left in solution, and the conductivity is zero.

After the end point, barium hydroxide solution is being added

Neutralization and salt formation

and the conductivity rises steeply, due to the mobility of the hydroxide ions.

Note: Since water shows very slight conductivity, the conductivity would not quite reach zero if a sensitive meter was used.

8.9 Heats of neutralization

Compare the reactions:

(a) $H^+Cl^-(aq) + Na^+OH^-(aq) \rightarrow Na^+Cl^-(aq) + H_2O$

Omitting spectator ions (Na^+ and Cl^- ions):

$H^+(aq) + OH^-(aq) \rightarrow H_2O$

(b) $(H^+)_2SO_4^{2-}(aq) + 2Na^+OH^-(aq) \rightarrow (Na^+)_2SO_4^{2-}(aq) + 2H_2O$

Omitting spectator ions (Na^+ and SO_4^{2-} ions)

$2H^+(aq) + 2OH^-(aq) \rightarrow 2H_2O$
or $H^+(aq) + OH^-(aq) \rightarrow H_2O$

That is, in acid/alkali neutralizations, the only energy change is due to the formation of water. So if equal numbers of hydrogen ($H^+(aq)$) and hydroxide ($OH^-(aq)$) ions are mixed, the heat change should be the same. This can be checked by experiment.

If the experiment was carried out using sulphuric acid solution and barium hydroxide solution, the rise in temperature will be higher than expected.

This is because there are two reactions taking place.

1 $H^+(aq) + OH^-(aq) \rightarrow H_2O$
2 $Ba^{2+}(aq) + SO_4^{2-}(aq) \rightarrow Ba^{2+}SO_4^{2-}(s)$

The second reaction accounts for the additional heat evolved.

8.10 Determining the concentration of a solution

These experiments are usually carried out using indicators, and not by conductivity experiments.

Examples

1 25 cm^3 of M hydrochloric acid neutralized 50 cm^3 of sodium hydroxide solution. Find the molarity of the sodium hydroxide.
(i) Balanced equation:

$$HCl(aq) + NaOH(aq) \longrightarrow NaCl(aq) + H_2O$$
$$\text{1 mole} \quad \text{1 mole}$$

$$\begin{array}{cc} \text{NaOH} & \text{HCL} \\ \text{n moles} \longleftrightarrow & \text{n moles} \end{array}$$

(ii) n = No. of moles NaOH   n = No. of moles HCl
    = (M × V)$_{NaOH}$ = (M × V)$_{HCl}$

(iii) ∴ (M × V)$_{NaOH}$ = (M × V)$_{HCl}$

$$\therefore \frac{M \times 50}{1000} = \frac{1 \times 25}{1000}$$

Simplify M × 50 = 25   ∴ M = $\frac{25}{50}$ = 0.5

∴ Molarity of NaOH = 0.5 M

2 100 cm^3 of $\frac{M}{10}$ sodium hydroxide (NaOH) was required to neutralize 25 cm^3 of sulphuric acid solution. What was the concentration of the acid?
(i) Balanced equation:

$$H_2SO_4 + 2NaOH \rightarrow Na_2SO_4 + 2H_2O$$
$$\text{mole} \qquad \text{2 moles}$$

$$\text{n moles} \longleftrightarrow \text{2n moles}$$

(ii) n = No. of moles H$_2$SO$_4$ 2n = No. of moles NaOH
    = (M × V)$_{H_2SO_4}$ = (M × V)$_{NaOH}$

$$\therefore n = \tfrac{1}{2}(M \times V)_{NaOH}$$

(iii) ∴ (M × V)$_{H_2SO_4}$ = $\tfrac{1}{2}$(M × V)$_{NaOH}$

$$M \times \frac{25}{1000} = \frac{1}{2} \times \frac{1}{10} \times \frac{100}{1000}$$

Simplify (1000 is common to both sides)

$$M \times 25 = \frac{1}{2} \times \frac{1}{10} \times 100 \quad \text{so } M = \frac{100}{2 \times 10 \times 25} = 0.2$$

Molarity of H$_2$SO$_4$ = 0.2 M.

Neutralization and salt formation

3 50 cm^3 of Phosphoric acid (H_3PO_4) is neutralized by 75 cm^3 of $\frac{M}{10}$, potassium hydroxide.

Find the concentration of the Phosphoric acid
(i) Balanced equation:

$H_3PO_4 + 3KOH \longrightarrow K_3PO_4 + 3H_2O$
1 mole   3 moles

We can now go straight to step (iii) questions 1 and 2

(M x V) acid = $\frac{1}{3}$ (M x V) alkali

Complete this question yourself.

Some revision questions

You should try to answer each question, and then check your answer by referring to the section indicated in brackets after the question.
1 Suggest how you might prepare samples of the following salts: (a) magnesium sulphate, (b) sodium nitrate, (c) copper carbonate. (Section 8.2 and 8.3.)
2 In an experiment you are asked to measure the conductivity of dilute sulphuric acid as sodium hydroxide solution is added to it.
   Write the equation for the reaction, and draw and explain the resulting conductivity graph you would expect.
(Section 8.4 and 8.6.)

3(a) 25 cm^3 of $\frac{M}{1}$ nitric acid neutralized 25 cm^3 of sodium hydroxide solution. Fine the molarity of the sodium hydroxide.

(b) 50 cm^3 of $\frac{M}{1}$ sodium hydroxide neutralized 50 cm^3 of sulphuric acid. Find the molarity of the acid
(Section 8.10)

9

Electrolysis of aqueous solutions

9.1 Copper (II) chloride solution

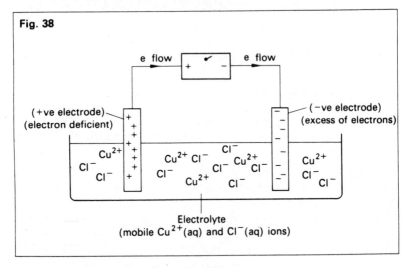

Fig. 38

At the positive electrode—Fig. 39

Negative chloride (Cl^-) ions are attracted to the positive electrode. Since the positive electrode is short of electrons, it pulls electrons from the chloride ions, and chlorine gas (Cl_2) is discharged.

$Cl^- \rightarrow Cl + e$
then $Cl + Cl \rightarrow Cl_2$
(or $2Cl^- \rightarrow Cl_2 + 2e$)   Oxidation

At the negative electrode—Fig. 40

Copper ions collect.
The negative electrode has an excess of electrons, so it pushes off 2e onto each copper ion, and copper metal is formed.

$Cu^{2+} + 2e \rightarrow Cu$ Reduction

Electrolysis of aqueous solutions

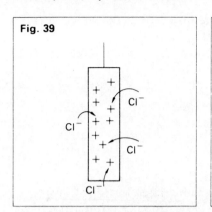

Fig. 39

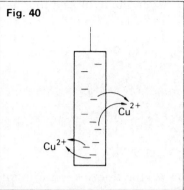

Fig. 40

9.2 Electrolysis of sodium halides

The table below shows the major products formed.

	Salt	−Electrode	+ Electrode
(a)	Sodium fluoride	Hydrogen	Oxygen
(b)	Sodium chloride	Hydrogen	Chlorine
(c)	Sodium bromide	Hydrogen	Bromine
(d)	Sodium iodide	Hydrogen	Iodine

We showed in section 7.8 that water contains a small number of ions:

$$H_2O(l) \rightleftharpoons H^+(aq) + OH^-(aq)$$

Therefore, in any solution, these ions must also be available for discharge.

We can use this fact to help explain the results of the electrolysis.

Reaction at Negative Electrode

In all cases, hydrogen gas is formed at the negative electrode, and *not* sodium.

The hydrogen ($H^+(aq)$) ions from the water accept electrons more readily than the sodium ions ($Na^+(aq)$).

Therefore hydrogen ions from the water are discharged as hydrogen gas:

$$2H^+(aq) + 2e \rightarrow H_2(g)$$

When the hydrogen ions are removed from the water, there will be an excess of hydroxide (OH⁻(aq)) ions.

The solution at the negative electrode becomes alkaline. This can be shown by an indicator solution.

Reaction at the Positive Electrode

(b) $2Cl^-(aq) \rightarrow Cl_2 + 2e$
(c) $2Br^-(aq) \rightarrow Br_2 + 2e$
(d) $2I^-(aq) \rightarrow I_2 + 2e$

In (a) the fluoride ion will not readily lose electrons to form fluorine, because fluorine has a very strong electron attracting power. Therefore, the fluoride ion is *not* discharged. Instead, hydroxide ions from the water discharge as oxygen gas.

$4OH^-(aq) \rightarrow 2H_2O + O_2(g) + 4e$

Since hydroxide ions from the water are discharged, an excess of hydrogen ions build up. Therefore the solution at the positive electrode becomes acidic (shown by the addition of an indicator).

9.3 Electrolysis of dilute sulphuric acid ('electrolysis of water')

(−)	(+)
$H^+(aq)$ ions collect	$SO_4^{2-}(aq)$ and $OH^-(aq)$ from water collect.
$H^+(aq)$ ions discharge	$OH^-(aq)$ ions from the water discharge in preference to the SO_4^{2-} ions.
$2H^+(aq) + 2e \rightarrow H_2(g)$	$4OH^-(aq) \rightarrow 2H_2O + O_2(g) + 4e$

Let us balance the reactions at the positive and negative electrodes, so that equal numbers of electrons are transferred.

$4H^+(aq) + 4e \rightarrow 2H_2$ $4OH^-(aq) \rightarrow 2H_2O + O_2 + 4e$

For every four hydrogen ions reduced at the negative electrode, four hydroxide ions are oxidised at the positive electrode.

The volume of hydrogen released at the negative electrode is twice the volume of oxygen released at the positive.

Electrolysis of aqueous solutions

In other words, water is being removed, and the electrolysis is equivalent to the electrolysis of water.

9.4 Preferential discharge of ions

Reaction at the negative electrode

The order below can be deduced from the data tables.

$Ag^+(aq) + e \rightarrow Ag$ }
$Cu^{2+}(aq) + 2e \rightarrow Cu$ } discharged from solution
$2H^+(aq) + 2e \rightarrow H_2$

H^+ ions from the water are discharged in preference to ions of active metals.
Ca^{2+}, Na^+, Li^+, K^+ etc., are *never* discharged from aqueous solution.

Reaction at the positive electrode

The order for chloride, bromide and iodide can be deduced from the data tables.

$2I^-(aq) \longrightarrow I_2 + 2e$ }
$2Br^-(aq) \longrightarrow Br_2 + 2e$ } most easily discharged
$2Cl^-(aq) \longrightarrow Cl_2 + 2e$ }
$4OH^-(aq) \longrightarrow 2H_2O + O_2 + 4e^-$

OH^- ions from the water are discharged in preference to other ions.
$F^-(aq)$, $SO_4^{2-}(aq)$, $NO_3^-(aq)$ are *never* discharged from aqueous solution. The equations for discharge can be found on pages 144 and 145.

Some revision questions
Copy and complete the following table, writing equations for the reactions at the electrodes, of the solutions listed below.

Solution of	− Electrode reaction	+ Electrode reaction
1 Copper (II) bromide, $CuBr_2$		
2 Silver nitrate, $AgNO_3$		
3 Sodium sulphate, Na_2SO_4		
4 Potassium iodide, KI		
5 Lithium fluoride LiF		
6 Copper (II) sulphate $CuSO_4$		

10

Sulphur and its compounds

10.1 Sulphuric acid

Sulphuric acid is one of the most widely used chemicals in any industrial society. Therefore, manufacture of this acid is of vital importance to the nation's economy.

Production of sulphuric acid must be based on the raw materials available. The main sources are:
1. The element sulphur – found uncombined in Texas and Sicily;
2. Iron pyrites ore (FeS_2) and other sulphide ores.

Both sulphur and iron pyrites can yield the gas sulphur dioxide. We will pause to consider this gas before continuing with sulphuric acid.

10.2 Sulphur dioxide (SO_2)

Tests for sulphur dioxide

(a) It turns *starch iodate* paper or solution a deep blue colour. This is a very sensitive test.
(b) It decolourises yellow-brown bromine water.
(c) It decolourises dark brown iodine solution.

Other materials react in the same way to tests (b) and (c), so these tests are not conclusive.

Obtaining sulphur dioxide from the raw materials

(a) Burning sulphur in air

$S + O_2 \longrightarrow SO_2$
(from air)

Sulphur and its compounds

(b) Heating iron pyrites, or any other sulphide ore, in air. Iron oxide and sulphur dioxide are formed.

$$4FeS_2 + 11O_2 \rightarrow 2Fe_2O_3 + 8SO_2$$

Properties of sulphur dioxide

1 It has a very unpleasant smell and irritates the throat and lungs.
2 It is easily liquefied — a mixture of solid carbon dioxide and alcohol will produce a temperature of about $-70°C$. This liquefies the sulphur dioxide.
It is therefore stored in cylinders as a liquid under pressure, rather than as a compressed gas.
3 It is very soluble in water. An acid solution (sulphurous acid) is formed.

$$H_2O + SO_2 \rightarrow 2H^+(aq) + \underbrace{SO_3{}^{2-}(aq)}_{\text{sulphite ion}}$$

$$\underbrace{\phantom{H_2O + SO_2 \rightarrow 2H^+(aq) + SO_3{}^{2-}(aq)}}_{\text{sulphurous acid}}$$

4 Although the gas does not burn, and does not usually support combustion (that is, other materials will not burn in sulphur dioxide), burning magnesium will continue to burn, forming sulphur and magnesium oxide.

$$2Mg + SO_2 \rightarrow 2MgO + S$$

Sulphur dioxide and pollution

Air pollution is a problem in all industrial areas, and one of the major pollutants is sulphur dioxide.
  Sulphur or sulphur compounds are found in many fuels — for example, heavy fuel oils, coal, etc. Therefore, any industry which burns these fuels produces sulphur dioxide. Coal or oil-burning power stations, and the steel industry all produce quantities of sulphur dioxide. The extremely poisonous nature of the gas makes it vital that the amount of sulphur dioxide is kept down to a tolerable level.

10.3 Reactions of the sulphite ion, SO_3^{2-}

In order to carry out experiments involving the sulphite ion, we must first be able to detect the presence of the *sulphate* ion, SO_4^{2-}.

Test for the sulphate ion in solution

If hydrochloric acid is added to the suspected sulphate, and then a solution containing barium (Ba^{2+}) ions is added (usually barium chloride), a white precipitate is produced if the sulphate ion is present.

e.g. $Na_2SO_4(aq) + BaCl_2(aq) \longrightarrow BaSO_4(s) + 2NaCl(aq)$

$SO_4^{2-}(aq) + Ba^{2+}(aq) \rightarrow Ba^{2+}SO_4^{2-}(s)$
white precipitate

The sulphite ion is a reducing agent

When reacting with materials which can be made to accept electrons (be reduced), the sulphite ion readily donates electrons, and in so doing is itself oxidised to the sulphate ion.

$SO_3^{2-} + H_2O \rightarrow SO_4^{2-} + 2H^+ + 2e$

(This equation can be found in the 'Electrochemical Series' table at the end of this book.)

The formation of the sulphate ion can be proved in the experiments below, by carrying out the test for the sulphate ion as detailed above.

(a) *With the halogens (Br_2 Cl_2 I_2)*

On adding bromine water to a solution containing sulphite ions, the red-brown bromine water is decolourised.

The bromine molecules have been reduced (gained electrons) to bromide ions.

$Br_2 + 2e \rightarrow 2Br^-$
brown   colourless

similarly,

$Cl_2 + 2e \rightarrow 2Cl^-$
green   colourless

$I_2 + 2e \rightarrow 2I^-$
brown   colourless

(b) *With iron (III) ions (Fe^{3+})*
The colour of the solution after mixing with sulphite ion solution changes from yellow to green, showing that Fe^{3+} ions have been reduced to Fe^{2+} ions. The presence of the ions can also be shown by the characteristic tests (see page 46).

Fe^{3+} + e → Fe^{2+}
yellow   green

Note: All the above reduction equations can be found in the data tables (page 144).

Proof of electron transfer in the above reactions (Fig. 41)
The solution in the right hand limb can be bromine water, iron (III) ion solution, silver ion solution, or any other material which will readily accept electrons.

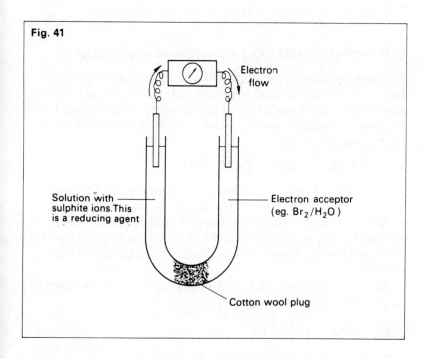

Fig. 41

Electrons flow from the sulphite ion solution to the bromine water (shown by the meter).

10.4 Uses of sulphites

(a) *Bleaching*. The sulphite ion can bleach paper or straw, and can safely bleach woollens and other delicate fabrics. This is again a reducing action, and when left exposed to air and sunlight they are re-oxidised and turn yellow.

(b) As a *disinfectant* and *preservative*. Due to its highly poisonous nature, trace amounts of sulphur dioxide can kill bacteria and micro-organisms, and it can therefore be used to preserve wines and foods.

(c) As a useful *source of sulphur dioxide*. Addition of moderately concentrated hydrochloric or sulphuric acid to any metal sulphite, and warming where necessary, yields sulphur dioxide gas.

$$SO_3^{2-} + 2H^+ \rightarrow H_2O + SO_2$$

For example: $Na_2SO_3 + 2HCl \rightarrow 2NaCl + H_2O + SO_2$

This reaction can be used as a test for the sulphite ion.

10.5 Sulphur trioxide (SO_3) and sulphuric acid (H_2SO_4)

The preparation of pure sulphuric acid by oxidation of the sulphite ion is too difficult to carry out on a large scale. Instead, sulphur dioxide is made to combine directly with oxygen, forming sulphur trioxide, which can be converted to sulphuric acid by dissolving in water.

The reaction: $2SO_2 + O_2 \rightarrow 2SO_3$

involves the collision of sulphur dioxide and oxygen molecules of sufficient energy to cause the change to take place.

At room temperature the energy of collision is not great enough for any noticeable reaction to take place. Therefore we must increase the speed of the molecules by raising the temperature.

However, when sulphur trioxide is heated, it breaks down to sulphur dioxide and oxygen.

$$2SO_3 \xrightarrow{heat} 2SO_2 + O_2$$

In other words, the reaction is reversible:

$$2SO_2 + O_2 \rightleftharpoons 2SO_3$$

Sulphur and its compounds

and at any temperature an *equilibrium* is reached where sulphur trioxide molecules are being broken down as fast as they are being made, and the percentage of sulphur trioxide formed depends on the temperature. In this case, the lower the temperature the higher the percentage of sulphur trioxide.

However, the lower the temperature, the lower the energy of collision of the molecules, and hence the longer it takes to reach equilibrium.

Our problem is:

High temperature – reaction fast, but % SO_3 formed too small.

Low temperature – % SO_3 high, but reaction too slow.

To speed up the rate at the lower temperature we use a catalyst (in this case *vanadium (V) oxide*, or *platinum* spread over mineral wool), which provides a big surface area for the molecules to react on. A temperature of about 450°C has been found to be a reasonable temperature for this reaction.

Preparation of sulphur trioxide and sulphuric acid

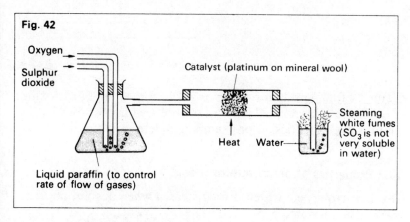

Fig. 42

$$2SO_2 + O_2 \xrightarrow[\text{heat (450°C)}]{\text{catalyst (platinum)}} 2SO_3$$

If the sulphur trioxide obtained is dissolved in water, sulphuric acid is obtained:

$$SO_3 + H_2O \rightarrow H_2SO_4$$

This process is important in industry, and is called the Contact Process. In modern plants, 99.5% conversion of sulphur dioxide to sulphur trioxide can be achieved.

The sulphur trioxide is not very soluble in water, and in industry it is normally dissolved in 98% sulphuric acid first, to which water is constantly being added to maintain the correct concentration.

10.6 Industrial preparation of sulphuric acid — The Contact Process — a simplified flow diagram

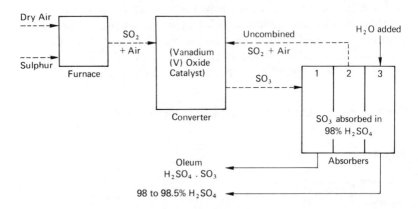

Oleum or fuming sulphuric acid, is a solution of sulphur trioxide in sulphuric acid. A small amount is produced in absorber 1 for use in some industries — for example, in detergent manufacture.

10.7 Properties of dilute sulphuric acid

This is a strong acid which is fully ionised when in solution in water. The ions are:

$2H^+(aq) + SO_4^{2-}(aq)$

It has all the properties associated with a strong acid:
(a) pH number is very low;
(b) it forms carbon dioxide gas when added to carbonates:

$$CuCO_3(s) + H_2SO_4(aq) \longrightarrow CuSO_4(aq) + H_2O + CO_2(g)$$
i.e. $CO_3^{2-}(s) + 2H^+(aq) \longrightarrow H_2O + CO_2(g)$

Sulphur and its compounds

(c) it reacts:
  (i) with an active metal
  $$Mg(s) + H_2SO_4(aq) \to MgSO_4(aq) + H_2(g)$$
  $$Mg(s) + 2H^+(aq) \to Mg^{2+}(aq) + H_2(g)$$
  (ii) by adding a base
  $$H^+(aq) + OH^-(aq) \to H_2O$$
  e.g. $H_2SO_4 + 2NaOH \to Na_2SO_4 + 2H_2O$
  or $2H^+(aq) + O^{2-}(s) \to H_2O$
  e.g. $H_2SO_4 + CuO \to CuSO_4 + H_2O$

10.8 Properties of concentrated sulphuric acid

1 The acid has a high boiling point, and this property helps us to prepare other acids with lower boiling points. For example, in the preparation of hydrochloric acid, sulphuric acid is added to a chloride.

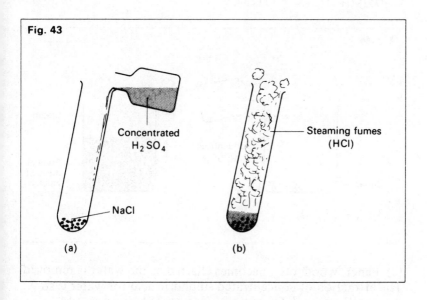

Fig. 43

$$H_2SO_4(l) + NaCl \longrightarrow NaHSO_4 + HCl(g)$$
i.e. $H_2SO_4(l) + Cl^- \longrightarrow HSO_4^- \quad + HCl(g)$
(hydrogen sulphate ion)

On dissolving the hydrogen chloride gas in water, hydrochloric acid is formed.

In the case of nitric acid, sulphuric acid is added to a nitrate:

$$H_2SO_4 (l) + KNO_3 \longrightarrow KHSO_4 + HNO_3 (g)$$
i.e. $H_2SO_4 + NO_3^- \longrightarrow HSO_4^- + HNO_3$

2 It has a great attraction for water, and will absorb water from the air. It is therefore a good drying agent for gases which do not react with it.

The affinity of sulphuric acid for water is so great that it will remove the elements of water from compounds which contain them:
(a) blue hydrated copper sulphate crystals turn white (become anhydrous) – the water of crystallisation is removed;
(b) sugar (sucrose) chars, and swells up as a black mass of carbon:

$$C_{12}H_{22}O_{11} \longrightarrow 12C + 11H_2O$$

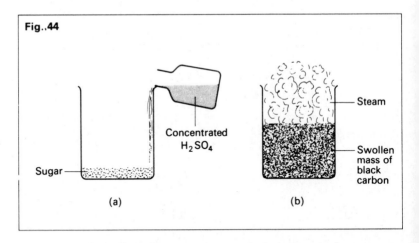

Fig..44

(a) Sugar / Concentrated H₂SO₄
(b) Steam / Swollen mass of black carbon

(c) paper, wood, etc., becomes charred as the water is removed. The attraction of concentrated sulphuric acid for water is so great, and so much heat is given out, that the acid should *always* be added to water and *never* water added to the acid. This is because the heat produced could turn drops of water added into steam, causing acid to spurt out.

Sulphur and its compounds

Concentrated Sulphuric acid is NOT highly ionised

(a) Blue litmus paper, one half wet, lies on top of some concentrated sulphuric acid. The wet half turns red immediately, the dry part stays blue.
(b) Pure, concentrated sulphuric acid is a poor conductor of electricity. It therefore contains only a few ions.

Concentrated sulphuric acid contains only a few ions, but when added to water it dissociates, forming a lot of $H^+(aq)$ ions. It is therefore a strong acid.

$$H_2SO_4(l) + H_2O(l) \rightarrow 2H^+(aq) + SO_4^{2-}(aq)$$

4. *Concentrated sulphuric acid is an oxidising agent*
(a) *Reaction with metals.* When hot, sulphur dioxide gas (SO_2) is formed. With copper and zinc.

$$\left. \begin{array}{l} Cu \longrightarrow Cu^{2+} + 2e \\ Zn \longrightarrow Zn^{2+} + 2e \end{array} \right\} \text{oxidation}$$

In the above cases, the sulphuric acid accepts electrons, and is reduced to sulphur dioxide gas.
Note: whereas copper will not react with dilute acids, it will react with concentrated sulphuric acid as above.
(b) *Reaction with non-metals.* With carbon the reaction is:

$$C + H_2SO_4 \rightarrow CO_2 + SO_2 + H_2O$$

Once again, the acid must be hot.

10.9 Uses of sulphuric acid

(a) In steel pickling – dipping steel into acid to clean it before treatment (for example before chrome plating).
(b) Manufacture of fertilisers – superphosphates and ammonium sulphate.
(c) Manufacture of dyes and paints.
(d) Manufacture of artificial fibres and plastics.
(e) Manufacture of soaps and detergents.

(f) Manufacture of other acids, e.g. hydrofluoric acid (HF) — used in glass etching.

Some revision questions

You should try to answer each question and then check your answer by referring to the section indicated in brackets after the question.

1 What acid forms when sulphur dioxide dissolves in water? Write its formula. (Section 10.2.)

2 (a) Give the test for the sulphate ion in solution.
 (b) What is meant when it is said that the sulphite ion is a reducing agent? Give two examples of its reducing action. You should use the Data Tables to help you write any equations. (Section 10.3.)

3 Give the conditions necessary and state the reactions involved in the preparation of sulphuric acid from sulphur dioxide. (Section 10.5.)

4 (a) How can sulphuric acid be used to prepare other acids? Give an example.
 (b) State some ways in which the properties of dilute and concentrated sulphuric acid differ.
 (Sections 10.7 and 10.8.)

11

Nitrogen and nitrogen compounds

11.1 Nitrogen and nitric acid

Nitrogen is the unreactive gas which occupies 4/5 of the air by volume.

The route to nitric acid formation can be similar to the route for sulphuric acid formation:

$$N_2 \xrightarrow{O_2} \text{Oxides of nitrogen} \xrightarrow{H_2O} HNO_3 \text{ (nitric acid)}$$

Nitrogen does not react readily with oxygen, but it can be made to combine if enough energy is used.

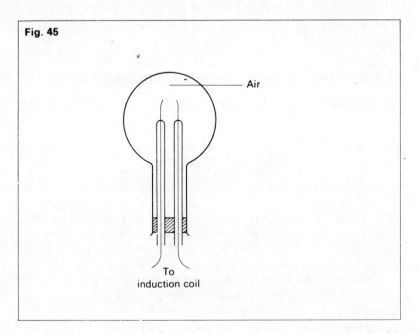

Fig. 45

Using an induction coil, sparks are passed through the oxygen/nitrogen (air) mixture in the flask (as in Fig. 45).
The brown gas, nitrogen dioxide, (NO_2) appears.
When water is added, pH paper turns red — nitric acid has been formed.
This method is *not economic, and so cannot be used.*
A similar reaction to the one just described takes place in air during lightning storms.

11.2 Nitrogen and hydrogen — ammonia (NH_3)

There are many *ammonium compounds* in the laboratory, e.g. ammonium chloride (NH_4Cl), and we can obtain ammonia gas from any ammonium compound by heating it with an alkali, e.g. sodium hydroxide, or calcium hydroxide.

$$NH_4Cl + NaOH \longrightarrow NaCl + NH_3 + H_2O$$
$$\text{i.e.} \quad NH_4^+ + OH^- \longrightarrow NH_3 + H_2O$$

The gas has a distinctive smell (it should be smelled *carefully*). Moist litmus or pH paper turns blue (hydroxide ions, OH^- are formed).

11.3 Ammonia and water — ammonia solution is a weak base

Ammonia is very soluble in water and forms an alkaline solution.

$$NH_3(g) + H_2O \longrightarrow NH_4^+(aq) + OH^-(aq)$$
alkaline due to presence of hydroxide ions.

However, if the conductivity of equal volumes of molar sodium hydroxide and molar ammonia solutions are compared, it is seen that the conductivity of the sodium hydroxide is high, while that of the ammonia solution is low.
Since both the sodium ions and the ammonium ions have nearly the same mobility and OH^- ions are common to both, the large difference in conductivity can only be explained by there being fewer ions present in the ammonia solution.

Sodium hydroxide is a *strong base*. When dissolved in water, it separates into free sodium ions and hydroxide ions.

$Na^+OH^-(s) + H_2O \rightarrow Na^+(aq) + OH^-(aq)$

An ammonia solution of the same concentration is a *weak base* and contains only a few ions. That is, at any one instant only a small number of ammonia molecules have combined with water molecules to form ions:

$NH_3 + H_2O \rightleftharpoons NH_4^+(aq) + OH^-(aq)$
most of the solution   only a few ions free
is in this form   at any instant

11.4 Combining nitrogen and hydrogen to form ammonia

In trying to form ammonia, we can try different energy sources.

Heating a mixture of nitrogen and hydrogen, or *sparking* the mixture does *not* yield ammonia.

In fact, if the experiment is tried in reverse, it is found that sparking dry ammonia will decompose it into the elements nitrogen and hydrogen (Fig. 46):

$2NH_3 \rightarrow N_2 + 3H_2$

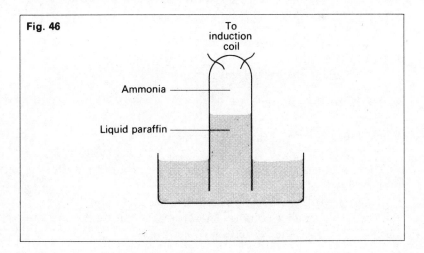

Fig. 46

The Haber Process

The reaction between nitrogen and hydrogen is reversible:

$$N_2 + 3H_2 \rightleftharpoons 2NH_3$$

Depending on conditions, the reaction will be driven towards the right or the left.
The reaction involves the collision of molecules of sufficient energy for the change to take place.

Effect of temperature
Since heating causes the ammonia to break down, then the temperature must be kept as low as possible.
But, if the temperature is too low, the reaction becomes very slow.
In other words:
    High temperature   —   reaction fast, but % NH_3 formed is too small.
    Low temperature   —   % NH_3 is high, but reaction too slow.

Effect of catalyst
To speed up the rate at the lower temperature, we use a catalyst (in this case, iron), which provides a large surface area for the molecules to react on.

Effect of pressure
If we look at the gas volumes involved:

$$\underline{N_2 + 3H_2} \rightleftharpoons 2NH_3$$

4 molecules $\rightleftharpoons$ 2 molecules

The 2 molecules of ammonia take up less room than the 4 molecules of mixture.
In other words, in forming ammonia we are trying to convert the gas into a smaller volume. Therefore, increasing the pressure should encourage the formation of ammonia.

Nitrogen and nitrogen compounds

A suitable balance of all three conditions must be found, and in practice, the conditions used are:

Temperature — about 500°C
Pressure — about 150 to 300 atmospheres
Catalyst — iron.

Although the percentage of ammonia produced is not high, the process is economical since any uncombined nitrogen and hydrogen can be recycled after removal of the ammonia.

For the industrial preparation of ammonia, see Appendix on page 139.

11.5 Oxidation of ammonia

Since the manufacture of nitric acid by oxidation of nitrogen is not economical, we need to find a method of oxidising the ammonia to the oxides of nitrogen, and hence to nitric acid.

Methods which might be attempted are:

1 Burning ammonia in oxygen

Ammonia will not burn readily in air, but it will do so in oxygen, the products being water and nitrogen.

$$2NH_3 + 3O_2 \longrightarrow N_2 + 3H_2O$$

Therefore, this is of *no value* in forming the oxides of nitrogen.

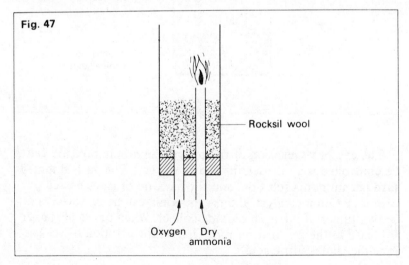

Fig. 47

2 Supplying oxygen from another compound

Ammonia is passed over hot copper (II) oxide.
Once again, nitrogen gas and water are formed.

$$2NH_3(g) + 3CuO(s) \longrightarrow N_2(g) + 3Cu(s) + 3H_2O(l)$$

3 Using a catalyst — The Ostwald Process

If oxygen is bubbled through ammonia solution, and a hot platinum wire lowered into the gas mixture, a reaction occurs, which is sometimes explosive. The reaction must be between the ammonia and oxygen, since oxygen bubbling through water shows no reaction.

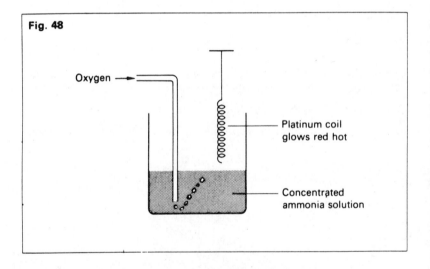

Fig. 48

The explosive reaction of the oxygen/ammonia mixture can be controlled if air is used instead of oxygen. The air is bubbled through ammonia solution, and the mixture of gases passed over a platinum catalyst. Excess ammonia can be removed, and brown fumes of nitrogen dioxide collect. When moist pH paper is placed in the gas, it turns red. That is, the solution of the gas is acidic, and is nitric acid.

Nitrogen and nitrogen compounds

The main reactions taking place are shown below. For simplicity, they are unbalanced.

a $NH_3 + O_2 \xrightarrow{\text{hot platinum catalyst}} NO + H_2O$
nitrogen monoxide
(nitric oxide)
(colourless gas)

b $NO + O_2 \longrightarrow NO_2$
from air in
the receiver
nitrogen dioxide
(brown gas)

c $NO_2 + H_2O \longrightarrow HNO_3$
nitric acid

This oxidation is used in industry, and is called the Ostwald Process. **For the industrial preparation and uses of nitric acid, see Appendix on page 140.**

11.6 Test for the nitrate ion, NO_3^-

(a) *Brown ring test*
Iron (II) sulphate solution is added to a solution of the suspected nitrate, and then a slow stream of concentrated sulphuric acid is poured down the side. This forms a layer at the bottom of the test-tube, and at the junction of the two layers a brown ring is obtained if the nitrate ion is present.

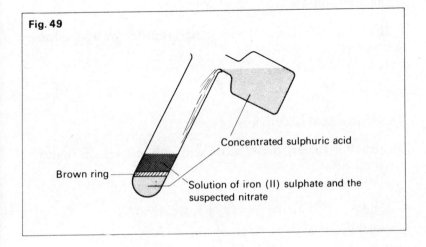

Fig. 49

Concentrated sulphuric acid

Brown ring

Solution of iron (II) sulphate and the suspected nitrate

(b) If copper and concentrated sulphuric acid is added to a nitrate (solid or solution), brown nitrogen dioxide gas is obtained.

11.7 Properties of nitric acid

Dilute nitric acid shows normal acid reactions

1 It turns pH paper red. It is a strong acid, and dissociates almost completely.

$$HNO_3(l) \xrightarrow{H_2O} H^+(aq) + NO_3^-(aq)$$

2 It is neutralized by an alkali (e.g. sodium hydroxide solution) or a base (e.g. copper oxide).

For example: $HNO_3(aq) + NaOH(aq) \rightarrow H_2O + NaNO_3(aq)$
or: $H^+(aq) + OH^-(aq) \rightarrow H_2O$

3 It reacts with carbonates, yielding carbon dioxide and water.

e.g.: $CuCO_3 + 2HNO_3 \rightarrow Cu(NO_3)_2 + H_2O + CO_2$
or: $CO_3^{2-} + 2H^+(aq) \rightarrow H_2O + CO_2(g)$

The metal nitrate is left in solution.

4 The action of nitric acid with metals is different from normal (see below).

11.8 Nitric Acid is an oxidising agent

(i) Reaction of very dilute nitric acid with magnesium (only) (Fig. 50(a))
Hydrogen is evolved:

$Mg(s) + 2HNO_3(aq) \rightarrow Mg(NO_3)_2(aq) + H_2(g)$
$Mg(s) \rightarrow Mg^{2+}(aq) + 2e$

Nitrogen and nitrogen compounds 97

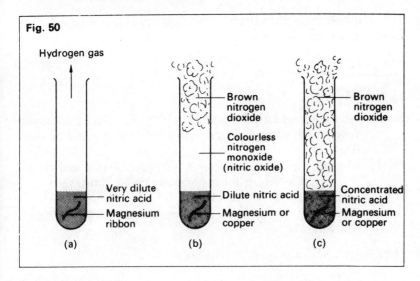

Fig. 50

(ii) Reaction of dilute nitric acid on magnesium or copper (Fig. 50(b)).

The magnesium or copper loses electrons (is oxidised).

$$Mg(s) \rightarrow Mg^{2+}(aq) + 2e \quad or \quad Cu(s) \rightarrow Cu^{2+}(aq) + 2e$$

The nitrate ion accepts these electrons and forms nitrogen monoxide (nitric oxide) gas (the nitrate ion is reduced).

$$\underbrace{NO_3^-(aq) + 4H^+(aq)}_{\text{from the acid}} + 3e \rightarrow NO + 2H_2O$$
nitrogen monoxide (nitric oxide)

When the nitrogen monoxide comes into contact with the air at the top of the tube, it reacts to form brown nitrogen dioxide gas.

(iii) Reaction of concentrated nitric acid with magnesium or copper (Fig. 50(c))

As before, the magnesium or copper loses electrons (is oxidised):

$$Mg(s) \rightarrow Mg^{2+}(aq) + 2e \quad or \quad Cu(s) \rightarrow Cu^{2+}(aq) + 2e$$

But this time, the nitrate ion accepts the electrons and forms brown nitrogen dioxide gas:

$$\underbrace{NO_3^-(aq) + 2H^+(aq)}_{\text{from acid}} + e \rightarrow NO_2(g) + H_2O$$

Note: All the ion-electron equations given here can be obtained from the Electrochemical Series Table at the end of this book.

Note: The gas which we have called nitrogen dioxide and given the formula NO_2, is more correctly called dinitrogen tetroxide, with formula N_2O_4.
As a result the equation for the reduction of the nitrate ion will now read:

$$2NO_3^-(aq) + 4H^+(aq) + 2e \rightarrow N_2O_4(g) + H_2O$$

This is the equation which you will find in the Electrochemical Series Table.

11.9 The importance of nitrogen compounds

Nitrogenous fertilisers

Plants and animals need to form nitrogen compounds (as proteins). Since nitrogen gas is very unreactive, it must be obtained in the form of nitrogen compounds.

Animals get the nitrogen by eating other animals or plants. Plants must obtain their nitrogen from the soil. We can help the plants to do this by adding fertilisers.

Fertilisers used are:

> ammonium sulphate — $(NH_4)_2SO_4$
> ammonium nitrate — NH_4NO_3
> ammonium hydrogen — $(NH_4)_2HPO_4$
>   phosphate

The Nitrogen Cycle

A few plants such as clover, bean and pea can use atmospheric nitrogen to build protein. These plants have *nodules* on the roots

Nitrogen and nitrogen compounds

which contain *nitrifying bacteria*, which convert atmospheric nitrogen into nitrates.

If all the nitrogen was put back into the soil as animal manure, compost, etc., and there was rotation of crops, there would be no need to use fertilisers. However, with most food consumed in cities, well away from the centre of growth, the Nitrogen Cycle is broken.

So the Nitrogen Cycle is 'assisted' by 'fixing' nitrogen by the Haber Process, and making fertilisers.

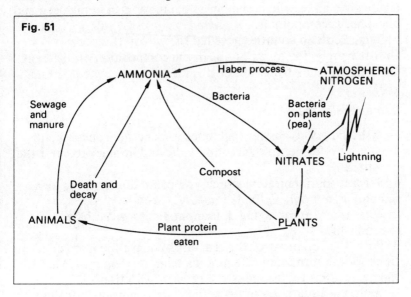

Fig. 51

The production of ammonia requires large amounts of fuel, and unless new fuel reserves are found, they cannot last for ever, so our problem has not been solved once and for all.

If the soil is acidic, denitrifying bacteria convert the nitrates back into nitrogen.

Lime can be used to reduce the acidity of the soil.

11.10 Conservation of sewage

Another important element for plant and animal growth is phosphorus.

Animal bones are composed of calcium phosphate, which is insoluble in water. Therefore, application of powdered bone

(bone meal) or crushed calcium phosphate rocks is very slow acting.

Soluble phosphates are:
    (a) ammonium hydrogen phosphate;
    (b) superphosphate (phosphate rocks treated with sulphuric acid form this fertiliser).

However, the bulk of phosphates and much of the nitrogen is lost to the sea as sewage, and since supplies of phosphate rocks are not unlimited, treatment of domestic sewage and dustbin refuse gives an excellent and cheap compost, rich in nitrogen and phosphorus compounds.

Some Scottish counties have led the way in this process. Unfortunately due to variations in the composition of different batches, the sale of this compost is difficult.

Some revision questions

You should try to answer each question and then check your answer by referring to the section indicated in brackets after the question.

1 Suggest any method by which you could quickly prepare a sample of ammonia in the laboratory. (Section 11.2.)

2 What is meant by saying that ammonia is a weak base? (Section 11.3.)

3 Write the equation for the reaction and give the main conditions for the formation of ammonia from its elements. What is the name given to this industrial process? (Section 11.4.)

4 Write the equations, and give the main conditions for the Ostwald Process for the oxidation of ammonia to nitric acid. (The equations need not be balanced.) (Section 11.5.)

5 List the main properties of dilute nitric acid, and indicate where they differ from other dilute acids. (Sections 11.7 and 11.8.)

6 Write down what you can about the nitrogen cycle, and the importance of nitrogen in everyday life. (Section 11.9.)

12

Fuels and related substances

12.1 The element carbon

Of the 100 or so elements on the Periodic Table, the element carbon is found in about $\frac{2}{3}$ of all known compounds. Carbon compounds are usually called *organic compounds*, and because they are so many and so varied they are studied in a separate branch of chemistry called organic chemistry.

First of all, let us look at the element carbon, a constituent in all these compounds.

12.2 Polymorphs of carbon

Carbon can exist in two different crystalline forms, having different physical properties. We call these forms *polymorphs* of carbon. **Polymorphs are different crystalline forms of the same element.**

The two polymorphs are: (a) diamond; (b) graphite.

Whereas diamond forms hard, clear, octahedral (eight-sided) crystals, and is a non-conductor of electricity, graphite forms soft plates and is a good conductor.

Explanation

In *diamond*, each carbon atom is linked covalently to four other carbon atoms, all through the crystal (Fig. 52) and a *macromolecule* is formed:

A macromolecule is a giant molecule, often consisting of more than 2000 atoms.

In *graphite*, each carbon atom has only three strong bonds. The fourth half filled orbital on each carbon only forms loose bonds with surrounding carbon atoms. The electrons in these orbitals are, therefore, fairly free, and graphite is a conductor.

Graphite only forming three strong bonds leads to the formation of flat plates (Fig. 53).

Fig. 52 Part of a diamond

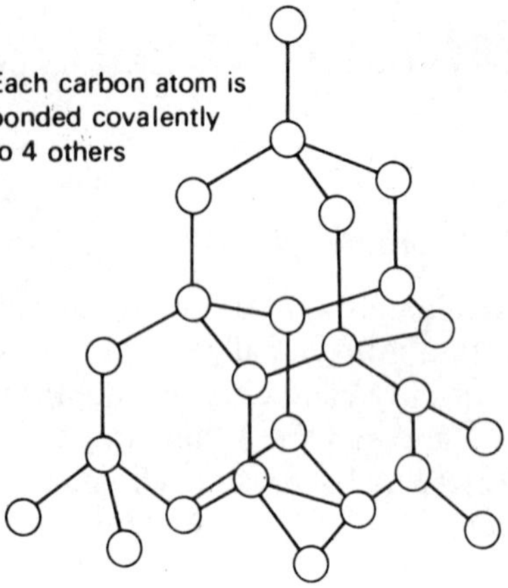

Each carbon atom is bonded covalently to 4 others

Fig. 53 Graphite

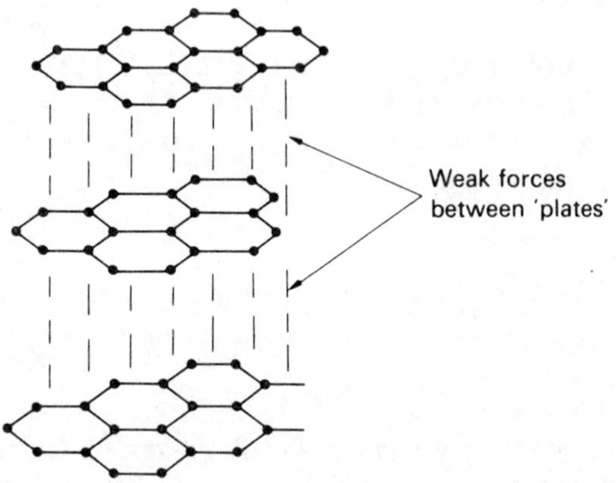

Weak forces between 'plates'

12.3 Carbon and carbon compounds as fuels

A fuel is a substance which can be used to obtain a supply of energy — usually in the form of heat.

Fuels and related substances

When carbon burns in oxygen, the gas carbon dioxide is formed, and a lot of heat energy is given out. Experiments can be carried out to show the value as fuels of petrol (sooty flame when burned in air, but very hot flame as fine droplets in a spray) and petrol vapour and coal gas (explosive mixtures with air).

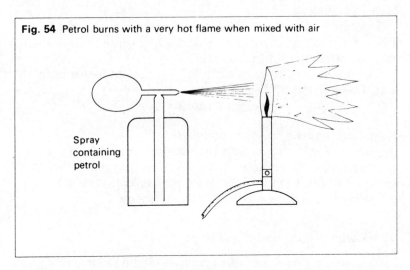

Fig. 54 Petrol burns with a very hot flame when mixed with air

In all cases, heat energy is given off, and carbon dioxide gas is formed.

Burning carbon (in a coal or coke fire)

A number of reactions can take place.
   (a) With plenty of air present:
$$C + O_2 \longrightarrow CO_2$$
                      carbon dioxide gas
   (b) With a limited supply of air
$$2C + O_2 \longrightarrow 2CO$$
                  carbon monoxide gas.

In a large fire, carbon monoxide is often formed, and this is seen burning with a blue flame at the surface of the fire, where more air is available.
Similar reactions take place when carbon compounds burn.

Carbon monoxide as a fuel

$$2CO + O_2 \longrightarrow 2CO_2$$

When carbon monoxide burns in air to give carbon dioxide, it releases heat energy in the process. This makes carbon monoxide a good fuel, and gases containing carbon monoxide are used as fuels in industry.

The carbon dioxide gas is identified by passing it through lime water (calcium hydroxide solution). A white precipitate of calcium carbonate is formed.

Carbon monoxide gas is identified by the fact that it burns with a blue flame, and forms carbon dioxide (tested with lime water – calcium hydroxide solution)

Poisonous nature of carbon monoxide
Carbon monoxide reacts with the haemoglobin in the blood, preventing the haemoglobin from absorbing oxygen. Hence if no oxygen can get into the blood from the lungs, the brain cells get starved of oxygen and death quickly follows.

12.4 Carbon forms chains and rings

Carbon atoms have 4 half-filled orbitals (and a valency of 4), and can form covalent bonds with other carbon atoms, making chains. Each carbon atom in the chain *must* have 4 bonds. The carbon atoms in the chain can also be joined to other atoms such as hydrogen, oxygen, nitrogen or chlorine.

A compound containing only the elements hydrogen and carbon is called a hydrocarbon.

There are several series of *hydrocarbons,* called *homologous series,* the simplest one being the **alkanes**.

	Expanded structural formula	Condensed structural formula
The 1st member is methane, CH_4	H–C(–H)(–H)–H (with H above and H below)	CH_4

Fuels and related substances

	Expanded structural formula	Condensed structural formula
The 2nd member is ethane, C_2H_6	H–CH$_2$–CH$_2$–H (with H's above and below each C)	$CH_3.CH_3$
the 3rd member is propane, C_3H_8	H–C–C–C–H (with H's on each C)	$CH_3.CH_2.CH_3$
the 4th member is butane, C_4H_{10}	H–C–C–C–C–H (with H's on each C)	$CH_3.CH_2.CH_2.CH_3$

Looking at the first 4 members of this homologous series of alkanes CH_4, C_2H_6, C_3H_8, C_4H_{10}, it can be seen that each fits the formula C_nH_{2n+2}, where n is the number of C atoms in the molecule.

C_nH_{2n+2} is the general formula for the alkane series.

If we look at butane (C_4H_{10}), we see that the 4 carbons and 10 hydrogens can join up in a different way, forming a branched chain:

```
          H
          |
       H—C—H
   H      |      H
   |      |      |
H—C ———— C ———— C—H
   |      |      |
   H      H      H
```

That is, $CH_3.\overset{CH_3}{\underset{\cdot}{CH}}.CH_3$ or, $(CH_3)_3.CH$

Molecules which have the same chemical formula, but different structural formulae, are called isomers.

The two molecules, C_4H_{10}, are isomers of butane. Try to draw the isomers of pentane (C_5H_{12}).

Where the carbons atoms of the hydrocarbon form a ring, you get a cycloalkane (Fig. 55).

Fig. 55 Cyclohexane

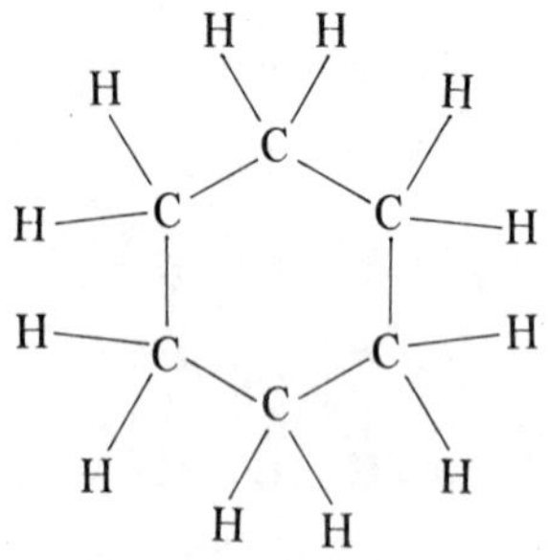

The molecular formula for this molecule is C_6H_{12}.

12.5 Reactions of hydrocarbons

(a) Combustion

The products of burning a hydrocarbon such as methane are carbon dioxide (lime water turns milky) and water (b.pt. = 100°C).

$$CH_4 + 2O_2 \longrightarrow CO_2 + 2H_2O$$

Hence, the carbon in the carbon dioxide, and the hydrogen in the water must have come from the fuel being burned.
Note: To prove that all the oxygen came from the air, and that there was none present initially in the fuel before burning, some other experiment would have to be done.
  Again, in a limited supply of air, the compounds are only partly burned, and carbon monoxide gas is formed.

(b) Substitution reactions

The alkanes are said to be *saturated* hydrocarbons (they have no double bonds), and because of this they tend to be unreactive, a hydrogen having to be pulled off before another atom can bond to the molecule.

Fuels and related substances

A slow *substitution* reaction takes place, for example, with bromine:

$$\underset{\begin{array}{c}|\ \ \ \ |\\H\ \ H\end{array}}{\overset{\begin{array}{c}H\ \ H\\|\ \ \ \ |\end{array}}{H-C-C-H}} + Br-Br \longrightarrow \underset{\begin{array}{c}|\ \ \ \ |\\H\ \ H\end{array}}{\overset{\begin{array}{c}H\ \ H\\|\ \ \ \ |\end{array}}{H-C-C-Br}} + H-Br \text{ (white fumes)}$$

Because the reaction is very slow, the red-brown bromine colour will persist for some time.

12.6 Separation of hydrocarbons — gas chromatography

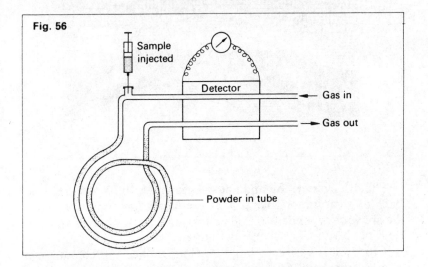

Fig. 56

Since hydrocarbons are all so similar chemically, those which cannot be separated by boiling point can be separated by chromatography. (See Fig. 56.)

A gas such as nitrogen is passed through a glass tube, and the detector adjusted to give no reading on the meter.

When the sample is injected (assuming it contains a mixture of hydrocarbons) each constituent will be absorbed differently by the powder and pass through the column at a different rate, so being separated.

Each time a hydrocarbon comes off the column, it causes a change in the detector and a reading on the meter is obtained.

12.7 The Oil Industry

Oil and natural gas are usually formed from the remains of tiny marine animals. The oil and gas are trapped underground by layers of rocks.

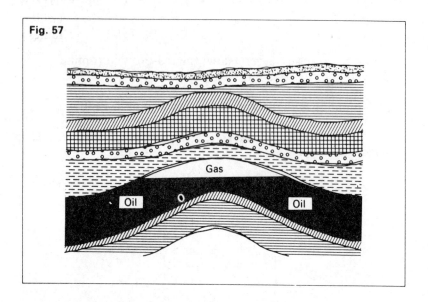

Fig. 57

The discovery of oil and natural gas under the North Sea has been of great importance to the economy of our country, since we are very dependent on oil and natural gas for power and heating. Natural gas is mainly methane, CH_4.

Crude oil is a dark, strongly smelling liquid. It is composed of a large number of different hydrocarbons, and the chemists' job is to separate them into useful constituents.

Fractional distillation of crude oil — refining

This is the process by which crude oil is separated into different components, using differences in boiling point. The molecules with a small number of carbon atoms have the lowest boiling points. All the oil is heated to a high temperature, and the gases produced rise up a tower, gradually cooling down. The longer chain molecules (with high boiling points) turn into the liquid state first and are separated off. The shorter chain molecules

Fuels and related substances

turn into the liquid state as they cool down further up the tower.

The laboratory experiment is different from the industrial process. The temperature of the oil sample is gradually increased, and the molecules with a small number of carbon atoms boil out of the crude oil mixture at low temperatures. As the temperature is raised, molecules with more carbon atoms boil off.

Product	Carbon atoms in chain
Gases	1–4
Petrol	4–12
Kerosene (paraffin)	9–16
Diesel oil	15–25
Lubricating oil	20–70
Bitumen	more than 70

12.8 Combustion of alkanes

If we examine the ease with which different oil fractions catch fire and burn, we find that the low boiling point liquids evaporate very quickly, and catch fire without the bunsen flame touching the liquid; they have a *low ignition temperature*. The ease of burning steadily decreases, the higher boiling liquids not burning even when a flame is directed on to them.

Hydrocarbons as fuels

We saw earlier that hydrocarbons are used as fuels, giving off a lot of heat (and light) energy when burned in air or oxygen. The burning of ethane is an example:

$$2C_2H_6 + 7O_2 \longrightarrow 4CO_2 + 6H_2O + \text{energy}$$

Where does the energy come from?
1 Breaking the C – C bonds, the C – H and the O – O bonds *requires* energy.
2 Making the C – O and the H – O bonds *releases* energy.
Much more energy is released in stage 2, than is required in stage 1.

The net result is that energy is released. *The reaction is exothermic.*

12.9 Putting the oil to use — thermal cracking and reforming

(a) Thermal cracking

The amount of heavy oils which are produced (about 50% of the crude petroleum) are far greater than the demand (about 1%).
 These molecules must be broken down into smaller, more useful units.
 If the molecules are heated strongly, they vibrate so violently that they are broken up. This is the process of *thermal cracking*.
 The cracking process may be speeded up by using a substance which gives a large, hot surface (for example, a dried clay). As a result the process is often called 'catalytic cracking' or 'cat cracking'.

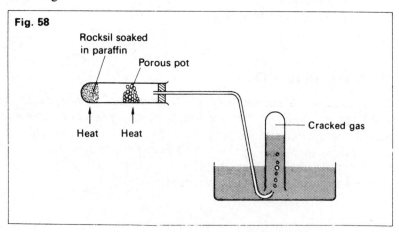

Fig. 58

If the cracked gas was shaken with bromine water, the colour of the bromine would disappear very quickly (compare the slow reaction with alkanes).
 Therefore, the cracked gases cannot all be alkanes.

What happens when molecules are cracked?

$$\begin{array}{c}\text{H H H H H H}\\|\ |\ |\ |\ |\ |\\\text{---C--C--C--C--C--C---}\\|\ |\ |\ |\ |\ |\\\text{H H H H H H}\end{array} \rightarrow \begin{array}{c}\text{H H H H H H}\\|\ |\ |\ |\ |\ |\\\text{---C--C}\cdot\ \cdot\text{C--C}\cdot\ \cdot\text{C--C---}\\|\ |\ |\ |\ |\ |\\\text{H H H H H H}\end{array}$$

Fuels and related substances

This leaves half-filled orbitals on the carbon atoms, and these molecules become stable by forming double bonds.

$$\begin{array}{cc} H & H \\ | & | \\ C= & C \\ | & | \\ H & H \end{array} \qquad \begin{array}{cc} H & H \\ | & | \\ C= & C \\ | & | \\ H & H \end{array}$$

Each time *one* bond is broken, two half-filled orbitals are produced and *one* double bond must be formed.

For example: $C_4H_{10} \rightarrow C_2H_6 + C_2H_4$
butane   ethane   ethene
                    (double bond)

These double bonded compounds, alkenes, give a fast reaction with bromine water, and are said to be *unsaturated*.

By careful control of the cracking process, ethene (C_2H_4) can be obtained as a major product.

(b) Catalytic reforming

Catalytic reforming involves changing the structure of molecules to give more useful molecules. This does *not* involve breaking carbon to carbon bonds.

Reforming is most commonly used to produce better petrol, since petrols containing cyclic or branched compounds give higher grade fuels. For example:

C_6H_{14} (hexane) $\longrightarrow$ C_6H_{12} (cyclohexane) + H_2.

Reforming can also be used to obtain alkenes from alkanes:

C_2H_6 (ethane) $\longrightarrow$ C_2H_4 (ethene) + H_2

$$\begin{array}{c} H \; H \\ | \; | \\ H-C-C-H \\ | \; | \\ H \; H \end{array} \longrightarrow \begin{array}{c} H \quad H \\ \backslash \;\; / \\ C=C \\ / \;\; \backslash \\ H \quad H \end{array} + H-H$$

These processes involve the use of a catalyst.

A homologous series of alkenes

A homologous series of alkenes exists — all having a carbon to carbon double bond.

$$\begin{array}{c}H\\|\\C\\|\\H\end{array}\!\!=\!\!\begin{array}{c}H\\|\\C\\|\\H\end{array}, \quad C_2H_4, \quad \text{ethene} ; \quad H-\begin{array}{c}H\\|\\C\\|\\H\end{array}-\begin{array}{c}H\\|\\C\\|\\H\end{array}\!\!=\!\!\begin{array}{c}H\\|\\C\\|\\H\end{array}, \quad C_3H_6, \text{propene}$$

The general formula of the alkene series is C_nH_{2n}

12.10 Reaction of bromine with saturated and unsaturated compounds

(a) *Saturated compounds*

$$H-\underset{\underset{H}{|}}{\overset{\overset{H}{|}}{C}}-\underset{\underset{H}{|}}{\overset{\overset{H}{|}}{C}}-H + Br-Br \longrightarrow H-\underset{\underset{H}{|}}{\overset{\overset{H}{|}}{C}}-\underset{\underset{H}{|}}{\overset{\overset{H}{|}}{C}}-Br + H-Br$$

This reaction is a *slow substitution*. The red-brown bromine colour persists for some time.

(b) *Unsaturated compounds*

$$\underset{\underset{H}{|}}{\overset{\overset{H}{|}}{C}}\!\!=\!\!\underset{\underset{H}{|}}{\overset{\overset{H}{|}}{C}} + Br-Br \longrightarrow Br-\underset{\underset{H}{|}}{\overset{\overset{H}{|}}{C}}-\underset{\underset{H}{|}}{\overset{\overset{H}{|}}{C}}-Br$$

The bromine has a strong attraction for electrons, and is attracted to the electron clouds of the double bonds. One of the bonds breaks, and a *fast addition reaction* takes place. The red-brown bromine colour disappears quickly.

This test will distinguish between saturated and unsaturated compounds.

12.11 Using the ethene — addition polymerisation

Once the ethene has been formed, one of the most important processes is to rebuild it into very long chain molecules, forming polythene (polyethene).

Fuels and related substances

Using an *initiator* to start the reaction, one of the bonds in the double bond splits, and the molecules link up to form a long chain of over 1000 carbon atoms

$$\underset{\text{ethene molecules}}{\overset{H}{\underset{H}{C}}=\overset{H}{\underset{H}{C}} + \overset{H}{\underset{H}{C}}=\overset{H}{\underset{H}{C}} + \overset{H}{\underset{H}{C}}=\overset{H}{\underset{H}{C}}} \longrightarrow \underset{\text{polythene}}{\cdots \overset{H}{\underset{H}{C}}-\overset{H}{\underset{H}{C}}-\overset{H}{\underset{H}{C}}-\overset{H}{\underset{H}{C}}-\overset{H}{\underset{H}{C}}-\overset{H}{\underset{H}{C}} \cdots}$$

Addition polymerisation takes place when molecules of an alkene (the monomer) which have a carbon to carbon double bond link up as a result of one of the bonds in the double bond splitting to form a long chain molecule (the polymer).

Ethene is the basis of all addition polymers. By replacing the H atoms of ethene by other atoms, or groups, different polymers with different properties can be obtained. For example:

P.V.C. (Polyvinyl chloride)

$$\underset{\substack{\text{monomer}\\\text{(vinyl chloride)}}}{\overset{Cl}{\underset{H}{C}}=\overset{H}{\underset{H}{C}}} \longrightarrow \underset{\substack{\text{polymer}\\\text{(polyvinyl chloride)}}}{\cdots \overset{Cl}{\underset{H}{C}}-\overset{H}{\underset{H}{C}}-\overset{Cl}{\underset{H}{C}}-\overset{H}{\underset{H}{C}}-\overset{Cl}{\underset{H}{C}}-\overset{H}{\underset{H}{C}} \cdots}$$

Polypropylene (polypropene)

$$\underset{\text{monomer}}{\overset{CH_3}{\underset{H}{C}}=\overset{H}{\underset{H}{C}}} \longrightarrow \underset{\text{polymer}}{\cdots \overset{CH_3}{\underset{H}{C}}-\overset{H}{\underset{H}{C}}-\overset{CH_3}{\underset{H}{C}}-\overset{H}{\underset{H}{C}}-\overset{CH_3}{\underset{H}{C}}-\overset{H}{\underset{H}{C}}-\overset{CH_3}{\underset{H}{C}}-\overset{H}{\underset{H}{C}} \cdots}$$

Teflon (P.T.F.E.)
(Polytetrafluoroethene)

$$\begin{matrix} F & F \\ | & | \\ C & = & C \\ | & | \\ F & F \end{matrix} \longrightarrow \quad \cdots C-C-C-C-C-C \cdots \text{(with F substituents)}$$

monomer　　　　　　　　polymer

Polystyrene

$$\begin{matrix} H & H \\ | & | \\ C & = & C \\ | & | \\ C_6H_5 & H \end{matrix} \longrightarrow \cdots C-C-C-C-C-C \cdots$$

Note: ⬡ represents a C_6H_5 'ring'.

12.12 Properties of polythene and polystyrene

	Melting	Burning	Insulating properties	Uses
Polythene	Melts and bubbles round edges	Burns quietly with blue/ yellow flame — tends to drip off.	Both are electrical and heat insulators.	Electric insulation, storage sacks, piping, kitchen ware etc.
Polystyrene	Harder to melt; bubbles round edges	Yellow flame, thick black smoke.		Toys, lampshades, guitars. As foam: ceiling tiles, packing materials, etc.

Some revision questions

You should try to answer each question and then check your answer by referring to the section indicated in brackets after the question.

1 (a) What is meant by (i) polymorph, (ii) macromolecule?
   (b) What are the polymorphs of carbon? How do their properties differ? (Section 12.2.)

Fuels and related substances

2 (a) What is a fuel?
(b) Carbon can burn to form different gases, depending on the amount of air present. What are the gases?
3 (a) What is meant by the following terms: (i) hydrocarbon; (ii) homologous series; (iii) structural formula; (iv) isomers?
(b) Draw the structural formulae for the isomers of butane. (Section 12.4.)
4 Describe the process of (i) fractional distillation, (ii) thermal cracking. (Sections 12.7 and 12.9.)
5 How could you distinguish between a saturated and an unsaturated hydrocarbon? (Section 12.10.)
6 (a) What is meant by 'addition polymerisation'?
(b) In addition polymerisation one *monomer* is involved. What is a monomer?
(c) Draw the structural formula of ethene.
(d) Draw the structural formula of polythene, showing how three ethene molecules have joined up together.
(Section 12.11.)

13
Foodstuffs and related substances

13.1 Carbohydrates

Carbohydrates are compounds containing carbon, hydrogen and oxygen with the hydrogen and oxygen in the proportions 2:1, as in water.

Carbohydrates are essentially chains of carbon atoms 'dressed' with H and OH groups:

$$\cdots C\underset{H}{\overset{OH}{|}} - C\underset{H}{\overset{OH}{|}} - C\underset{H}{\overset{OH}{|}} - C\underset{H}{\overset{OH}{|}} \cdots$$

Some proof of the composition is obtained from the fact that:
(a) combustion in oxygen produces carbon dioxide and water;
(b) when concentrated sulphuric acid is added to a carbohydrate (e.g. sugar) the elements of water are removed, leaving only carbon.

Comparison of starch and glucose — two carbohydrates (Fig. 59)

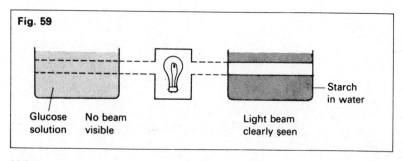

Foodstuffs and related substances 117

Glucose forms a true solution with water, while starch forms a colloid (its molecules must be very big — big enough to scatter light). In other words, glucose appears to be composed of small molecules, while starch has very large molecules.

Classes of carbohydrate

1 *Monosaccharides (simple sugars)*. For example: glucose and fructose. They have the formula $C_6H_{12}O_6$ (a chain of six carbon atoms).
2 *Disaccharides* (complex sugars). For example: maltose and sucrose. They have the formula $C_{12}H_{22}O_{11}$ (a chain of 12 carbon atoms).
3 *Polysaccharides* (non-sugars). Very long chain molecules of formula $(C_6H_{10}O_5)_n$ where n = approximately 300 for starch or glycogen and approximately 3000 for cellulose.

13.2 Tests for saccharides

1 *All sugars except sucrose*
React with Fehling's solution, or Benedict's reagent to give an orange-red precipitate. These sugars are called *reducing sugars*.
2 *Starch*
Reacts with iodine, giving a blue-black colour.
If the above tests are carried out on plants and foodstuffs, the presence of starch and reducing sugars is shown in many of them.

13.3 Connection between starch and glucose (hydrolysis of starch)

In the presence of acid, starch molecules can be broken down, by combining with water, to a reducing sugar. (The reducing sugar formed is, in fact, glucose.) Hydrochloric acid is used to help the reaction.

$$(C_6H_{10}O_5)_n + n\, H_2O \rightarrow n\, C_6H_{12}O_6$$
starch　　　　　　　　　　glucose

The starch has been *hydrolysed*.
Hydrolysis is the process by which organic molecules are broken down by addition of water.

This suggests that starch is made up by the polymerisation of a large number of glucose molecules (water being lost in the process):

A polymerisation where water is lost in the process, is called a condensation polymerisation.

$$n\ C_6H_{12}O_6 \underset{\text{hydrolysis}}{\overset{\text{condensation}}{\rightleftharpoons}} (C_6H_{10}O_5)_n + n\ H_2O$$
glucose $\qquad\qquad\qquad\quad$ starch

Condensation is the process by which organic molecules join together, by losing water.

13.4 Hydrolysis of starch in our bodies

Starch is broken down to the reducing sugar, maltose, by saliva. This can be shown by experiment, using a sample of your own saliva.

The breakdown is carried out by a catalyst called an *enzyme*.

An enzyme is a big molecule with a definite shape which allows it to carry out a particular reaction (to build up, or break down a molecule). An enzyme is a specialised catalyst which can only do one job.

In the breakdown of starch by saliva, the enzyme *amylase* (sometimes called *ptyalin*) in the saliva assists the breakdown process.

13.5 Carbohydrates as energy sources

When carbohydrates are burned in oxygen they produce a great amount of energy. A mixture of air and a fine powder such as flour, is explosive. Carbohydrates are used as energy sources in, for example, the burning of paper, or wood in a fire.

Within the body, the same amount of energy is released but, of course, very much more slowly at the normal body temperature of 37°C. We eat carbohydrates, which we store and 'burn' up at the required time to give us energy, the process being called *respiration*. Plants must manufacture their own carbohydrates, and do this by the process of *photosynthesis*, energy being taken in during the process.

13.6 Photosynthesis

Carbohydrates are made in plants from carbon dioxide (in the air) and water (from the soil). Light energy is supplied from the sun, and chlorophyll (the green colouring matter) is required to help the process. Since light energy is required, the process is called photosynthesis.

$$6 CO_2 + 6 H_2O + \text{light energy} \xrightarrow[\text{(in presence of chlorophyll)}]{\text{photosynthesis}} C_6H_{12}O_6 + 6 O_2 \text{ monosaccharide (e.g. glucose)}$$

Oxygen gas is released into the air in this process. The glucose formed is then polymerised to starch and cellulose, and the starch is stored until required.

Photosynthesis is the process by which plants manufacture carbohydrates from carbon dioxide and water, using light energy in the presence of chlorophyll. Oxygen is released in the process.

13.7 Respiration

All living cells, whether plants or animals, require energy, and this energy is obtained by the 'combustion' of carbohydrates. The starch is broken down (*hydrolysed*) to glucose, and the glucose (using oxygen) is broken down into carbon dioxide and water, releasing energy.

This process is *respiration*, and is the reverse of photosynthesis.

$$C_6H_{12}O_6 + 6 O_2 \underset{\text{photosynthesis}}{\overset{\text{respiration}}{\rightleftharpoons}} 6 CO_2 + 6 H_2O + \text{Energy}$$

Respiration is the process by which plants and animals obtain a supply of energy by breaking down carbohydrates (using oxygen) into carbon dioxide and water, the energy being released in the process.

Respiration in Animals

This is summarized in the chart below.

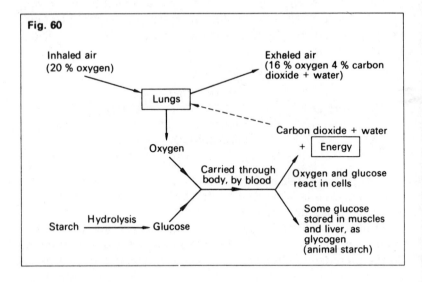

Fig. 60

13.8 The Carbon Cycle

This is best summarized in the chart below.

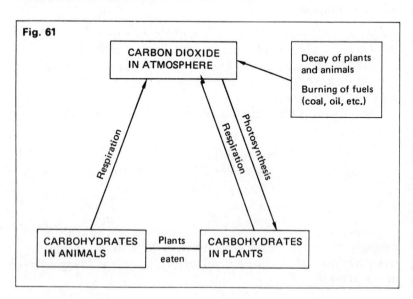

Fig. 61

Foodstuffs and related substances

13.9 Disaccharides

Sucrose is the only non-reducing sugar (the only sugar which doesn't react with Fehling's or Benedict's solutions).

Hydrolysis of sucrose

When sucrose is hydrolysed using acid, reducing sugars are obtained.

$$C_{12}H_{22}O_{11} + H_2O \longrightarrow 2C_6H_{12}O_6$$
Sucrose   Reducing Sugars

The presence of the reducing sugars is shown by Fehling's test. (In this test, the deep blue colour of Fehling's solution changes to green, and then a reddish-brown precipitate of copper(I) oxide forms.)

Identifying the products of hydrolysis of sucrose — chromatography

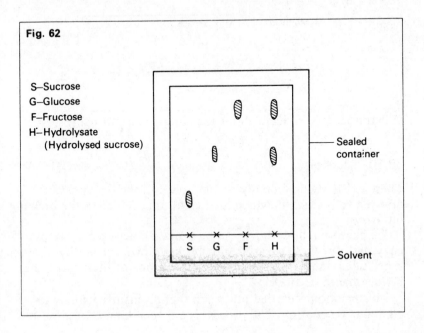

Fig. 62

S—Sucrose
G—Glucose
F—Fructose
H—Hydrolysate
(Hydrolysed sucrose)

Known sugars and the hydrolysate (hydrolysed sucrose) are spotted on the paper in a line at the points marked 'X' (Fig. 62). As the liquid soaks up the paper, it washes the carbohydrates along at different rates.

When the liquid has moved up the paper, the paper is dried and the position of the spots shown by using a spray which reacts with the sugars.

In this case the hydrolysed sucrose is shown to contain glucose and fructose, since it separates into two spots which moved the same distance as the glucose and fructose samples.

This shows that: Glucose + Fructose $\underset{\text{Hydrolysis}}{\overset{\text{Condensation}}{\rightleftharpoons}}$ Sucrose.

Hydrolysis of Maltose shows the following results:

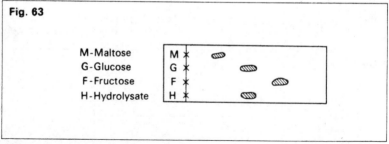

Fig. 63

M - Maltose
G - Glucose
F - Fructose
H - Hydrolysate

What does this indicate about the composition of maltose?

13.10 Alcohols

There exists a whole family of alcohols, but the one most often referred to is *ethanol*, which is of great importance to the brewing industry.

We have seen how starch can be broken down to glucose by enzymes. But these sugars can be broken down further by another series of enzymes into simpler molecules, one of these molecules being ethanol (C_2H_5OH).

This reaction is carried out when there is insufficient air for complete respiration. The process is a special kind of respiration.

Foodstuffs and related substances

13.11 The Brewing industry

The process can be summarized as taking place in the following steps.
1 Barley shoots are ground up to obtain the enzyme, amylase (sometimes called diastase).
2 This is added to starchy material, and

$$\text{starch} \xrightarrow{\text{amylase}} \text{maltose}$$

3 Yeast is now added and the yeast supplies enzymes, called maltase, and zymase.
These catalyse the following reactions:

(a) $C_{12}H_{22}O_{11} + H_2O \xrightarrow{\text{maltase}} 2\ C_6H_{12}O_6$
    maltose   glucose
(b) $C_6H_{12}O_6 \xrightarrow{\text{zymase}} 2C_2H_5OH + 2CO_2$
    glucose   ethanol

4 The result is a dilute ethanol solution, which can be fractionally distilled to yield a solution containing about 95% ethanol.

Note: You have now come across two other names for *amylase*.
    1 In saliva, it used to be called *ptyalin*.
    2 In plants, it used to be called *diastase*.
    The names ptyalin and diastase are still used occasionally.

Fermentation of glucose into ethanol

The reaction is similar to the final stages of the brewing industry reactions above, and can readily be carried out in the laboratory.

13.12 Polar nature of ethanol

```
    H   H                                      H   H
    |   |                                      |   |    δ−  δ+
H — C — C — H      One 'H' replaced by 'OH'  H—C — C — O — H
    |   |                                      |   |
    H   H                                      H   H
```

ethane (non polar)   ethanol (polar molecule)

The oxygen in the ethanol, having a strong attraction for electrons, makes the molecule polar and therefore gives ethanol different properties from ethane.

13.13 Properties of ethanol

(a) Ethanol is completely soluble in water.
(b) Its pH = 7, and it is a non-conductor of electricity.
(c) It burns, giving off carbon dioxide and water.

$$C_2H_5OH + 3O_2 \rightarrow 2CO_2 + 3H_2O$$

(Note: This experiment would only prove C and H to be present in ethanol. Since oxygen was used in burning, all the oxygen in the products could have come from the air.)
(d) Ethanol may be oxidised to ethanoic acid (acetic acid). If wine is not stored properly it goes sour, due to the ethanol being oxidised to ethanoic acid. Industry uses large amounts of ethanoic acid, and it can be made from ethanol chemically. Ethanol vapour reacts with hot copper (II) oxide, copper being formed and ethanoic acid vapour being produced (shown by pH indicator solution turning red).

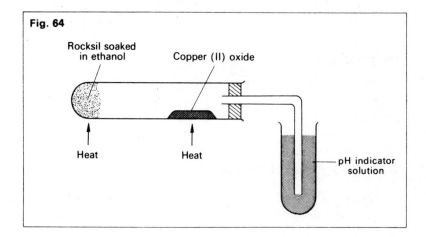

Fig. 64

$$2CuO + CH_3CH_2OH \rightarrow 2Cu + \underset{\text{ethanoic acid}}{CH_3COOH} + H_2O$$

Foodstuffs and related substances

13.14 Ethanoic acid CH_3COOH

Structure

```
    H    O
    |    ‖
H — C — C
    |    \
    H     O — H
```

Properties

(a) Soluble in water, having a pH less than 7. Therefore it is acidic. The pH of dilute solutions is usually found around 3; therefore, it is a *weak acid*.
(b) Conductivity is very low compared with acids like hydrochloric acid. This confirms that it is a weak acid.
(c) It shows normal acid properties; for example, it reacts with magnesium to produce hydrogen.

$$2CH_3COOH + Mg \longrightarrow (CH_3COO)_2Mg + H_2$$

Ethanoic acid — a weak acid

When dissolved in water, only a few of its molecules dissociate into ions. Therefore, at any one instant, only a small proportion of acid is present as ions.

$CH_3COOH + H_2O \rightleftharpoons CH_3COO^-(aq) + H^+(aq)$
most of the solution   only a few ions free at
is in this form   any instant

13.15 A series of acids

There is a whole series of acids. Each one has a similar formula and name based on the alkane with the corresponding number of carbon atoms; the name ends in '-oic acid'.

For example: HCOOH   methanoic acid
              $CH_3.CH_2.COOH$ propanoic acid

13.16 Reaction of acids with alcohols — esters

Esters are formed as a result of a condensation reaction (removal of water) between an acid and an alcohol. Usually concentrated

sulphuric acid is used to assist the removal of water, and the mixture is heated on a water bath for some time.
Two examples are given below:

$$CH_3.COOH + HOCH_2.CH_3$$
ethanoic acid   ethanol

condensation ↓ (H_2O removed)

$$CH_3.COOCH_2.CH_3 + H_2O$$
ethyl ethanoate (an ester)

$$CH_3.COOH + HOCH_3$$
ethanoic acid   methanol

condensation ↓ (H_2O removed)

$$CH_3.COO.CH_3 + H_2O$$
methyl ethanoate (an ester)

Naming of Esters

The name of an ester depends on the acid and alcohol from which it is prepared.
For example:—

$CH_3 \cdot CH_2 \cdot COO$	$CH_2 \cdot CH_3$
Name based on 'parent' acid — propanoic acid.	Name based on 'parent' alcohol — ethanol.
Therefore named *propanoate*	Named *ethyl*

Therefore name of ester: *ethyl propanoate*.

Hydrolysis of esters is the reverse of *condensation,* the ester being broken up into the acid and alcohol.

Uses of esters

(1) Nail varnish, and nail varnish remover.
(2) Car paint thinners.
(3) Artificial flavourings.
(4) Perfumes and cosmetics (esters have a 'fruity' smell).

13.17 Fats and oils

All fats and oils are of vegetable or animal origin. They are *esters* of the polyalcohol (an alcohol with a number of OH groups) called glycerol.

The acids forming the esters are often the liquid oleic acid or the solid stearic acid.

$$C_{17}H_{35} \cdot COOH \ + \ -\overset{|}{\underset{|}{C}}-OH \ \xrightarrow{condensation} \ C_{17}H_{35} \cdot COO\overset{|}{\underset{|}{C}}- \ + \ H_2O$$

Stearic acid   part of glycerol molecule   a fat

$$C_{17}H_{33} \cdot COOH \ + \ -\overset{|}{\underset{|}{C}}-OH \ \xrightarrow{condensation} \ C_{17}H_{33} \cdot COO\overset{|}{\underset{|}{C}}- \ + \ H_2O$$

oleic acid   an oil

Fats and oils as energy sources

Fats are important energy sources, although it is doubtful if they are essential for the release of energy in animals.

Fats give out more energy than the same weight of carbohydrates, and for this reason people in cold climates eat much more fat than others in warmer countries.

When fats are digested, the first breakdown results in the formation of glycerol and the acid. These are transported through the body, where they either break further, releasing energy, or they recombine to form fat.

Oils are unsaturated

Oils in animals and plants (for example, whale oil, coconut oil, olive oil) are liquid at ordinary temperatures. They are formed from unsaturated acids, and will decolourise bromine water. (See test for unsaturated compounds, section 12.10.)

128 'O' grade chemistry

13.18 Margarine manufacture

The oils are 'hardened' by adding hydrogen across the double bonds, using a catalyst. The 'hardened' material is then mixed with the correct amount of unsaturated oil until the right consistency is achieved.

13.19 Soap is manufactured from fats and oils

When fats or oils are hydrolysed by boiling with a solution of an alkali (for example, sodium hydroxide), glycerol and an acid, such as stearic acid, are produced. The acid is immediately neutralized by the alkali to form the salt, sodium stearate (a soap).

$$\text{Fat + sodium hydroxide} \xrightarrow{\text{hydrolysis and neutralization}} \text{sodium stearate (soap) + glycerol + water}$$

The soap and glycerol are separated by 'salting' — adding sodium chloride solution.

The soap molecule consists of two parts: the long chain $C_{17}H_{35}$, which is covalent, and the remainder, $COO^- Na^+$, which is ionic.

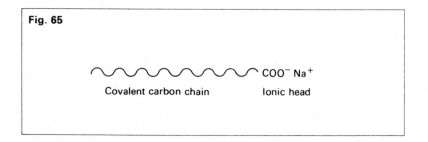

Fig. 65

Covalent carbon chain $COO^- Na^+$ Ionic head

13.20 The cleansing (detergent) action of soap

A soap molecule has a covalent tail (insoluble in water, soluble in oil) and an ionic head (soluble in water, insoluble in oil). (Fig. 65.)

Oil and water don't mix. If soap molecules are added they line up as in Fig. 66(a). When the mixture is shaken soap molecules form round each oil droplet as in Fig. 66(b).

The positive sodium ions move into the water, leaving a series

Foodstuffs and related substances

of negative charges as in Fig. 66(c). This prevents the droplets coming together, and so forms an *emulsion*.

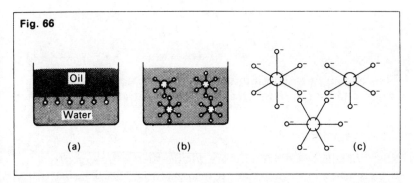

Fig. 66

Washing the dishes

In removing grease from a plate, soap acts in exactly the same way, forming an emulsion of the grease in water.

13.21 The action of soap on hard water

In some areas of Britain, the water in the reservoirs contains sufficient calcium ions (Ca^{2+}) or magnesium ions (Mg^{2+}) to cause a problem. This water is called *hard water,* since it forms a scum with soap instead of forming a lather. The scum is insoluble calcium stearate, or calcium oleate.

Sodium stearate + calcium ions →Calcium stearate + sodium ions
(soap)   (insoluble scum)

Solving the problem of hard water

1 The calcium or magnesium ions can be removed chemically. This can be expensive.
2 **Soapless detergents** may be used. Chemists have produced a molecule which acts like a soap but does not react with calcium or magnesium ions. The petroleum industry can supply an abundance of long chain hydrocarbons, and a long hydrocarbon chain, ending in a ring of six carbon atoms is chosen. When this is treated with oleum (fuming sulphuric acid) it takes on a suitable head group, which, when neutralized, gives the detergent (Fig. 67).

Fig. 67

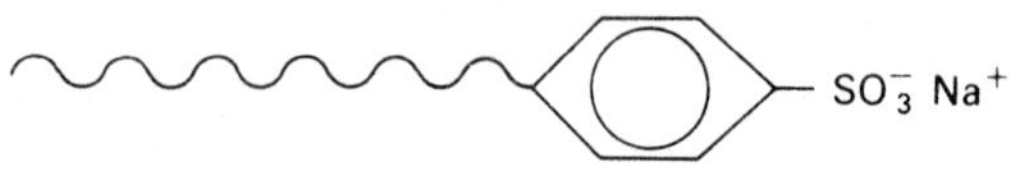

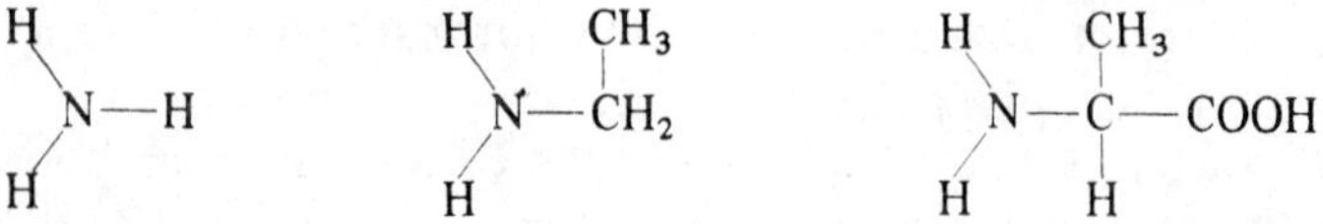

~~~~~⟨⬡⟩—SO_3^- Na^+

The calcium or magnesium salts of this detergent are soluble in water, and therefore the detergent gives a lather with hard water.

What is a detergent?

Soaps or soapless detergents are *all* classified as detergents, because they have the same cleansing action. (See 13.20.)

13.22 Proteins

Proteins are another group of long chain molecules. They are the 'building material' of nature.

Common protein foods are meat, eggs, fish, etc. Other protein materials are skin, hair, fingernails, wool, gelatin.

Composition of proteins

When a protein is heated with solid sodium hydroxide, a gas comes off which smells like ammonia, and turns pH paper blue (that is, the gas is alkaline).

Thus, a protein is related in some way to ammonia.

13.23 Amino acids, the building bricks

ammonia

H
 \
  N—H
 /
H

a base
(alkaline in water)

amine

H CH_3
 \ |
  N—CH_2
 /
H

a base

amino acid

H CH_3
 \ |
  N—C—COOH
 / |
H   H

basic   acid
group   group

Foodstuffs and related substances

There are about 24 amino acids, which make up all natural proteins by different combinations. Amino acids only differ in the length of the carbon chain, and may be represented:—

$$\begin{array}{c} H \quad R_1 \\ \diagdown \mid \\ N-C-COOH \\ \diagup \mid \\ H \quad H \end{array} \qquad \begin{array}{c} H \quad R_2 \\ \diagdown \mid \\ N-C-COOH \\ \diagup \mid \\ H \quad H \end{array} \qquad \begin{array}{c} H \quad R_3 \\ \diagdown \mid \\ N-C-COOH \\ \diagup \mid \\ H \quad H \end{array}$$

(R_1, R_2 and R_3 are carbon chains of different lengths.)

Amino acids condense to form proteins

H_2N--------COOH H_2N--------COOH

+ H_2O ↑ ↓ –H_2O
(hydrolysis)   (condensation)

H_2N--------CO—NH--------COOH

The CO—NH link formed in the condensation is called the Peptide Link.

A protein molecule is formed when a large number of the amino acid molecules join up together.

13.24 Hydrolysis of proteins and detection of the amino acids formed

Proteins can be hydrolysed (broken down) in a test tube by heating with hydrochloric acid.

The amino acids can be separated, and detected by paper chromatography.

13.25 Digestion of proteins

The enzyme *pepsin* digests (hydrolyses) the protein into its amino acids. These are absorbed in the blood and transported to part of the body which is growing, or in need of repair. There, the amino acids are reassembled (condensed) into a different order, to make the proteins required for body tissue.

Some revision questions

You should try to answer each question, and then check your answer by referring to the section indicated in brackets after the question.

1 Name the three classes of carbohydrate and give the general formula for each class. (Section 13.1.)

2 (a) What is the test for a reducing sugar? Which sugar will *hot* give a positive test?
  (b) What is the test for starch?
  (c) How could you show that starch was built up from reducing sugars?
  (d) What is meant by the word *hydrolysis*?
  (Sections 13.1 to 13.4.)

3 Describe the processes of *photosynthesis* and *respiration*. (Section 13.6 and 13.7.)

4 How could you distinguish between glucose and fructose? (Section 13.9.)

5 Describe how you could prepare a sample of ethanol starting from glucose. (Section 13.11.)

6 Give an equation, and describe how you could oxidise ethanol to ethanoic acid. (Section 13.13.)

7 What is an ester, and how could you make one? Give an example. (Section 13.16.)

8 (a) Describe how fat can be converted into soap.
  (b) How can soap be used to remove grease from clothing?
  (c) What is a soapless detergent? Why is it more useful in hard water areas?
  (Sections 13.19 to 13.21.)

9 (a) What is the 'building brick' of a protein?
  (b) Show how any three of these 'building bricks' combine together to form a protein.
  (Section 13.23.)

14

Macromolecules

A macromolecule is a very big molecule in which a very large number of the atoms are joined by covalent bonds.

Polymerisation is the process where a number of simple molecules called monomers join up together, usually forming a long chain. A macromolecule is formed.

There are two ways by which small molecules can be linked with each other to give larger ones.

(1) Addition polymerisation.

Here only one type of monomer is used. The monomer is based on ethene ($CH_2 = CH_2$), and they join together by breaking one of the carbon–carbon bonds, and linking head to tail.

This was covered in Section 12.11.

e.g.

$$\begin{array}{c} H \quad H \\ | \quad | \\ C=C \\ | \quad | \\ H \quad H \end{array} + \begin{array}{c} H \quad H \\ | \quad | \\ C=C \\ | \quad | \\ H \quad H \end{array} + \begin{array}{c} H \quad H \\ | \quad | \\ C=C \\ | \quad | \\ H \quad H \end{array} \rightarrow$$

$$\ldots\ldots \begin{array}{c} H \quad H \quad H \quad H \quad H \quad H \\ | \quad | \quad | \quad | \quad | \quad | \\ C-C-C-C-C-C \\ | \quad | \quad | \quad | \quad | \quad | \\ H \quad H \quad H \quad H \quad H \quad H \end{array} \ldots\ldots$$

Rubber – a natural addition polymer
The chains in rubber are similar to polythene, but double bonds are still present. These double bonds can be attacked by atmospheric oxygen and the rubber becomes perished.
Vulcanising – Sulphur atoms link between chains of rubber molecules, cutting down the double bonds and at the same time making it less flexible.

(2) Condensation polymerisation

The monomers join together with the loss of water.
Natural polymers are usually formed by condensation reactions – examples are starch and cellulose (condensation of glucose) and proteins (condensation of amino acids).

In *some* natural condensation polymerisations, one monomer only is involved, for example glucose polymerising to starch. In others (e.g. protein formation) a number of monomers are involved.

Synthetic condensation polymerisation

Two different monomers are involved, and water is eliminated between the molecules:

$$HO\text{-}\boxed{A}\text{-}OH \quad H\text{-}\boxed{B}\text{-}H \quad HO\text{-}\boxed{A}\text{-}OH \quad H\text{-}\boxed{B}\text{-}H$$
$$\downarrow -H_2O$$
$$\text{---}\boxed{A}\text{-}\boxed{B}\text{-}\boxed{A}\text{-}\boxed{B}\text{---}$$

14.1 Synthetic condensation polymers

These can be derived from two main sources:
(a) converting existing polymers such as cellulose (cotton wool) into more suitable structures such as rayon;
(b) by building up new polymers from a variety of monomers, (for example terylene, urea-formaldehyde).
  Nylon is made from two monomers:
(1) a string of carbon atoms (X) with an acid group at each end

$$HOOC\text{-}\boxed{X}\text{-}COOH$$

(2) a string of carbon atoms (Y) with an amine ($-NH_2$) group at each end

$$H_2N\text{-}\boxed{Y}\text{-}NH_2$$

The NH_2 group (as in amino acids) is related to ammonia, and has basic properties.
  Condensation occurs between the acidic and basic groups.

$$HOOC\text{-}\boxed{X}\text{-}COOH \quad H_2N\text{-}\boxed{Y}\text{-}NH_2 \quad HOOC\text{-}\boxed{X}\text{-}COOH \quad H_2N\text{-}\boxed{Y}\text{-}NH_2$$
$$\downarrow -H_2O$$
$$\text{---}\boxed{X}\text{-}CO\cdot NH\text{-}\boxed{Y}\text{-}NH\cdot CO\text{-}\boxed{X}\text{-}CO\cdot NH\text{-}\boxed{Y}\text{---}$$

Macromolecules

In the laboratory, it is easy to make nylon if we have an acid chloride group (—COCl) at each end of the molecule 'X', instead of an acid group.

An experiment in which nylon is formed is shown in Fig. 68.

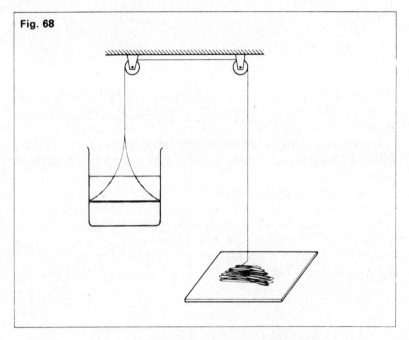

Fig. 68

The top layer is a diamine ($H_2 N$—$\boxed{Y}$—NH_2) dissolved in sodium hydroxide solution. The bottom layer is a diacid chloride (ClOC—$\boxed{X}$—COCl) dissolved in tetrachloromethane.

Various kinds of Nylon can be obtained, depending on the length of carbon chains.

Animal fibres, such as hair, wool and silk are proteins. The artificial fibre, nylon, has also the properties of protein — it gives off ammonia with sodium hydroxide pellets.

14.2 Thermoplastics and thermosets

Nylon can be melted and cooled back to its original form. It is said to be a *thermoplastic polymer*. Thermoplastics can be melted and reshaped over and over again. Examples are nylon, terylene, perspex, polythene, P.V.C., polystyrene, P.T.F.E.

They are usually composed of long chain molecules, which when cold, are tangled up together.

Fig. 69

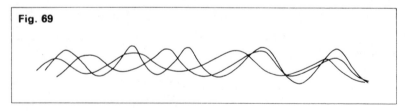

Heating untangles the chains, due to the increased vibrations of the molecules, and the polymer becomes more pliable and melts.

Urea-formaldehyde can be shaped and baked to form a hard solid which does not melt easily. It is said to be a *thermosetting* polymer.

Thermosetting polymers are polymers which harden on heating and do not melt on reheating. Other thermosetting polymers are epoxy resins (such as Araldite) and melamine-formaldehyde (Formica). Heating a thermosetting polymer causes the chains to cross-link, making the polymer rigid, and difficult to melt (Fig. 70).

Fig. 70

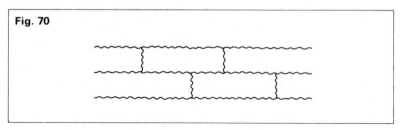

14.3 Identification of polymers by burning

Most plastics burn in different manners and it is possible to identify some of them by their burning properties.

Substance	Burning property
Nylon	Blue flame with yellow tip and no smoke. Smells of burning wool. The material drips and the flame tends to go out.
Cellulose Acetate	Yellow flame, some light smoke. There is a smell of vinegar and often spurting.
Polythene	Burns quietly, blue/yellow flame – tends to drip.
Polystyrene	Yellow flame, thick black smoke – drips.
PTFE	Does not burn. Feels greasy when cold.

14.4 Silicones

Silicon is an element in group 4 of the Periodic Table and, like carbon, can form 4 covalent bonds.

Silicon atoms link alternately with oxygen atoms, and the other silicon bonds are attached to methyl groups ($-CH_3$).

$$-O-\underset{\underset{CH_3}{|}}{\overset{\overset{CH_3}{|}}{Si}}-O-\underset{\underset{CH_3}{|}}{\overset{\overset{CH_3}{|}}{Si}}-O-\underset{\underset{CH_3}{|}}{\overset{\overset{CH_3}{|}}{Si}}-O-\underset{\underset{CH_3}{|}}{\overset{\overset{CH_3}{|}}{Si}}-O-$$

Varying the length of the Si−O− chain, and introducing cross linking between chains, gives a great variety of silicones, from oily liquids through waxes and rubbery plastics to hard solids.

Groups other than the CH_3- group are sometimes introduced to give even more variation in properties.

Properties and uses of silicones

1 Their water repellant and anti-stick properties are used in polishes and non-stick linings.
2 They are good insulators of heat and electricity, and can withstand high temperatures. Silicone rubbers can withstand heat and petrol, where ordinary rubbers and plastics do not. Silicones can also be used in electrical insulation of motors which have to withstand heat and water.

To prevent short circuits, the electronics industry insulates micro circuits by coating them in silicones.

Some revision questions

You should try to answer each question, and then check your answer by referring to the section indicated in brackets after each question.
1 What is meant by *addition polymerisation*? Give an example. (See introductory section.)
2 What is meant by *condensation polymerisation*? Give an example. (See introductory section, and 14.1.)
3 What is (i) a thermoplastic, (ii) a thermoset? (Section 14.2).
4 Draw part of a silicone chain. Why are silicones valuable? (Section 14.4).

Appendix: Salt preparations

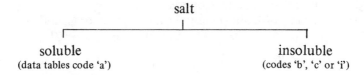

Insoluble

Prepared by precipitation:

Solution of required metal ion   +   Solution of non-metal
(the nitrate is always soluble)   ion (the sodium or
                                        potassium compound
                                        is always soluble)

Example: Silver chloride AgCl

$AgNO_3$ (aq) + NaCl (aq) ⟶ AgCl (s) + $NaNO_3$ (aq)

Soluble

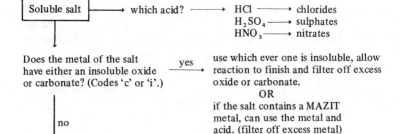

Does the metal of the salt   yes   use which ever one is insoluble, allow
have either an insoluble oxide ⟶   reaction to finish and filter off excess
or carbonate? (Codes 'c' or 'i'.)   oxide or carbonate.
                                                  OR
                                     if the salt contains a MAZIT
            no   metal, can use the metal and
                                     acid. (filter off excess metal)

Use the metal hydroxide (alkali) with
acid, using indicator to show the
end-point of the reaction.

Examples

(a) $Cu(NO_3)_2$
    either HNO_3 (aq) + $CuCO_3$ (s)
    or HNO_3 (aq) + CuO(s)

(b) $NaNO_3$
    HNO_3 (aq) + NaOH(aq)

Appendix: Industrial preparation of ammonia

Source of nitrogen and hydrogen — a flow diagram

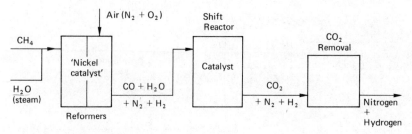

The nitrogen is obtained from the air.

Hydrogen is now mainly obtained from methane, by reacting it with steam and air, using a nickel catalyst:

(i) $CH_4 + H_2O \longrightarrow CO + 3H_2$
(ii) $2CH_4 + 3O_2 \longrightarrow 2CO + 4H_2O$ } Reactions in the Reformers

Reaction (ii) above takes place because air must be introduced to supply the nitrogen.

The carbon monoxide is converted to carbon dioxide in the "Shift Reactor" by reacting with steam and a catalyst. More hydrogen is produced.

$CO + H_2O \longrightarrow CO_2 + H_2$

The Haber Process — a flow diagram

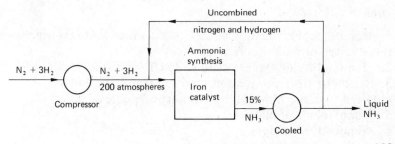

Appendix: Nitric acid manufacture

The Ostwald Process — a flow diagram

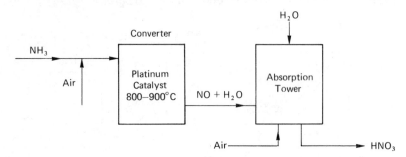

The main reactions taking place are shown below. For simplicity, they are unbalanced.

(a) In the converter:

$$NH_3 + O_2 \xrightarrow{\text{hot platinum catalyst}} NO + H_2O$$

nitrogen monoxide
(nitric oxide)
(colourless gas)

(b) In the absorption tower:

$$NO + O_2 \longrightarrow NO_2$$
from air

nitrogen dioxide
(brown gas)

$$\text{and } NO_2 + H_2O \longrightarrow HNO_3$$

nitric acid

Uses of Nitric Acid

1. Manufacture of fertilisers. 80% of the nitric acid manufactured is converted into ammonium nitrate.
2. Manufacture of explosives (e.g. T.N.T. or nitroglycerine).
3. Synthetic fibre production used to manufacture materials required to make fibres such as nylon and polyesters.
4. Manufacture of plastics.
5. Manufacture of dyes.

Appendix: Data tables

Ionisation Energies of Selected Atoms

Element	Symbol	Atomic number	Ionisation Energies in kJ mol^{-1}			
			First	Second	Third	Fourth
Hydrogen	H	1	1320	—	—	—
Helium	He	2	2380	5260	—	—
Lithium	Li	3	526	7310	11800	—
Beryllium	Be	4	905	1770	14800	—
Boron	B	5	807	2440	3660	25000
Carbon	C	6	1090	2360	4640	6220
Nitrogen	N	7	1410	2860	4580	7470
Oxygen	O	8	1320	3400	5320	7470
Fluorine	F	9	1690	3380	6060	8410
Neon	Ne	10	2090	3960	6140	9360
Sodium	Na	11	502	4560	6920	9540
Magnesium	Mg	12	744	1460	7750	10500
Aluminium	Al	13	584	1830	2760	11600
Silicon	Si	14	792	1590	3250	4350
Phosphorus	P	15	1020	1920	2930	4950
Sulphur	S	16	1010	2260	3380	4560
Chlorine	Cl	17	1260	2310	3840	5160
Argon	Ar	18	1530	2670	3950	5770
Potassium	K	19	425	3060	4440	5880
Calcium	Ca	20	596	1160	4930	6470
Scandium	Sc	21	637	1250	2410	7130
Titanium	Ti	22	664	1320	2670	4170
Vanadium	V	23	656	1430	2850	4600
Chromium	Cr	24	659	1600	3000	4800
Manganese	Mn	25	723	1520	3270	5000
Iron	Fe	26	766	1570	2970	5480
Cobalt	Co	27	764	1660	3250	—
Nickel	Ni	28	743	1770	3410	5400
Copper	Cu	29	751	1970	3570	5700
Zinc	Zn	30	913	1740	3850	5990
Arsenic	As	33	953	1800	2750	4830
Bromine	Br	35	1150	2100	3480	4560
Rubidium	Rb	37	409	2670	3880	—
Strontium	Sr	38	556	1080	4120	5500
Silver	Ag	47	737	2080	3380	—
Tin	Sn	50	715	1420	2960	3930
Antimony	Sb	51	816	1610	2460	4260
Iodine	I	53	1020	1850	2040	—
Caesium	Cs	55	382	2440	—	—
Barium	Ba	56	509	979	3420	—
Gold	Au	79	896	1990	—	—
Lead	Pb	82	722	1460	3100	4080
Bismuth	Bi	83	705	1620	2470	4370

Notes

The first ionisation energy for an element E refers to the reaction $E(g) \rightarrow E^+(g) + e$.
The second ionisation energy refers to $E^+(g) \rightarrow E^{2+}(g) + e$.

"Atomic Weights", "Atomic" Radii, and Electronegativities of Selected Elements

Element Symbol	[1]"Atomic Weight"	Covalent Radii (single) nm	Metallic Radii nm	[2,3,4]Ionic	Radii nm	Electronegativity (Pauling scale)
H	1	0.037	—	H^-	0.154	2.1
He	4	—	—			—
Li	7	0.123	0.152	Li^+	0.068	1.0
Be	9	0.089	0.112	Be^{2+}	0.035	1.5
B	11	0.080	—	B^{3+}	0.023	2.0
C	12	0.077	—	C^{4+}	0.016	2.5
N	14	0.074	—	N^{3-}	0.171	3.0
O	16	0.074	—	O^{2-}	0.132	3.5
F	19	0.072	—	F^-	0.133	4.0
Ne	20	—	—			—
Na	23	0.157	0.186	Na^+	0.097	0.9
Mg	24	0.136	0.160	Mg^{2+}	0.066	1.2
Al	27	0.125	0.143	Al^{3+}	0.051	1.5
Si	28	0.117	—	Si^{4+}	0.042	1.8
P	31	0.110	—	P^{3-}	0.212	2.1
S	32	0.104	—	S^{2-}	0.184	2.5
Cl	35.5	0.099	—	Cl^-	0.181	3.0
Ar	40	—	—			—
K	39	0.203	0.231	K^+	0.133	0.8
Ca	40	0.174	0.197	Ca^{2+}	0.099	1.0
Sc	45	0.144	0.161	Sc^{3+}	0.073	1.3
Ti	48	0.132	0.145	Ti^{2+}	0.094	1.5
V	51	0.122	0.131	V^{2+}	0.088	1.6
Cr	52	0.117	0.125	Cr^{2+}	0.089	1.6
				Cr^{3+}	0.063	
Mn	55	0.117	0.129	Mn^{2+}	0.080	1.5
Fe	56	0.116	0.124	Fe^{2+}	0.074	1.8
				Fe^{3+}	0.064	
Co	59	0.116	0.125	Co^{2+}	0.072	1.8
Ni	59	0.115	0.125	Ni^{2+}	0.069	1.8
Cu	64	0.117	0.128	Cu^{2+}	0.072	1.9
Zn	65	0.125	0.133	Zn^{2+}	0.074	1.6
As	75	0.121	—	As^{3-}	0.222	2.0
Br	80	0.114	—	Br^-	0.196	2.8
Rb	86	0.216	0.248	Rb^+	0.147	0.8
Sr	88	0.191	0.215	Sr^{2+}	0.112	1.0
Ag	108	0.134	0.144	Ag^+	0.126	1.9
Sn	119	0.140	0.140	Sn^{2+}	0.093	1.8
Sb	122	0.141	—	Sb^{3-}	0.245	1.9
I	127	0.133	—	I^-	0.220	2.5
Cs	133	0.235	0.265	Cs^+	0.167	0.7
Ba	137	0.198	0.217	Ba^{2+}	0.134	0.9
Au	197	0.134	0.144	Au^+	0.137	2.4
				Au^{3+}	0.085	
Pb	207	0.154	0.175	Pb^{2+}	0.120	1.8
				Pb^{4+}	0.084	
Bi	209	0.152	0.155	Bi^{3+}	0.096	1.9

Notes

[1] The "atomic weight" is also known as "relative atomic mass".

[2] Many elements form more than one ion. In the table two ions are shown for chromium, iron, gold, and lead.

[3] The ionic species shown may not be the usual form, e.g. H^-, Si^{4+}.

[4] Values for "atomic" radii are often quoted in Angstrom units (Å). 1 Å = 0.1 nm. 1 nm = 10^{-9} m.

Electrochemical Series—Standard Reduction Potentials

Reaction		Potential volts
$Li^+(aq) + e$	$\rightarrow Li(s)$	-3.02
$Rb^+(aq) + e$	$\rightarrow Rb(s)$	-2.92
$K^+(aq) + e$	$\rightarrow K(s)$	-2.92
$Sr^{2+}(aq) + 2e$	$\rightarrow Sr(s)$	-2.89
$Ca^{2+}(aq) + 2e$	$\rightarrow Ca(s)$	-2.76
$Na^+(aq) + e$	$\rightarrow Na(s)$	-2.71
$Mg^{2+}(aq) + 2e$	$\rightarrow Mg(s)$	-2.37
$Al^{3+}(aq) + 3e$	$\rightarrow Al(s)$	-1.70
$Zn^{2+}(aq) + 2e$	$\rightarrow Zn(s)$	-0.76
$Cr^{3+}(aq) + 3e$	$\rightarrow Cr(s)$	-0.74
$Fe^{2+}(aq) + 2e$	$\rightarrow Fe(s)$	-0.41
$Cr^{3+}(aq) + e$	$\rightarrow Cr^{2+}(aq)$	-0.41
$Ni^{2+}(aq) + 2e$	$\rightarrow Ni(s)$	-0.23
$Sn^{2+}(aq) + 2e$	$\rightarrow Sn(s)$	-0.14
$Pb^{2+}(aq) + 2e$	$\rightarrow Pb(s)$	-0.13
$Fe^{3+}(aq) + 3e$	$\rightarrow Fe(s)$	-0.04
$2H^+(aq) + 2e$	$\rightarrow H_2(g)$	0.00
$S(s) + 2H^+(aq) + 2e$	$\rightarrow H_2S(aq)$	0.14
$Sn^{4+}(aq) + 2e$	$\rightarrow Sn^{2+}(aq)$	0.15
$Cu^{2+}(aq) + e$	$\rightarrow Cu^+(aq)$	0.16
$SO_4^{2-}(aq) + 2H^+(aq) + 2e$	$\rightarrow SO_3^{2-}(aq) + H_2O$	0.20
$Hg_2Cl_2(s) + 2e$	$\rightarrow 2Hg(l) + 2Cl^-(aq)$	0.27
$Cu^{2+}(aq) + 2e$	$\rightarrow Cu(s)$	0.34
$I_2(s) + 2e$	$\rightarrow 2I^-(aq)$	0.54
$Fe^{3+}(aq) + e$	$\rightarrow Fe^{2+}(aq)$	0.77
$Ag^+(aq) + e$	$\rightarrow Ag(s)$	0.80
$2NO_3^-(aq) + 4H^+(aq) + 2e$	$\rightarrow N_2O_4(g) + 2H_2O$	0.81
$Hg^{2+}(aq) + 2e$	$\rightarrow Hg(l)$	0.85
$NO_3^-(aq) + 4H^+(aq) + 3e$	$\rightarrow NO(g) + 2H_2O$	0.96
$Br_2(l) + 2e$	$\rightarrow 2Br^-(aq)$	1.07
$MnO_2(s) + 4H^+(aq) + 2e$	$\rightarrow Mn^{2+}(aq) + 2H_2O$	1.21
$O_2(g) + 4H^+(aq) + 4e$	$\rightarrow 2H_2O$	1.23
$Cr_2O_7^{2-}(aq) + 14H^+(aq) + 6e$	$\rightarrow 2Cr^{3+}(aq) + 7H_2O$	1.33
$Cl_2(aq) + 2e$	$\rightarrow 2Cl^-(aq)$	1.36
$MnO_4^-(aq) + 8H^+(aq) + 5e$	$\rightarrow Mn^{2+}(aq) + 4H_2O$	1.49
$F_2(g) + 2e$	$\rightarrow 2F^-(aq)$	2.85

Note

The data given above are reduction potentials applicable to standard state conditions. This should be kept in mind when applying the data to ionic solutions.

Appendix

Data Applicable to the "Electrolysis of Water"

Reduction reactions at the negative electrode
$2H_2O + 2e \rightarrow H_2(g) + 2OH^-(aq)$ -0.83 V
$2H^+(aq) + 2e \rightarrow H_2(g)$ 0.00 V

Oxidation reactions at the positive electrode
$2H_2O \rightarrow O_2(g) + 4H^+(aq) + 4e$ -1.23 V
$4OH^-(aq) \rightarrow 2H_2O + O_2(g) + 4e$ -0.40 V

Note
The selection of the appropriate half-cell reactions depends on the nature of the ionic solution being considered.

Formulae of Selected Ions containing more than one Atom

Ion	Formula	Ion	Formula	Ion	Formula
Ammonium	NH_4^+	Hydrogen-sulphate	HSO_4^-	Permanganate	MnO_4^-
Carbonate	CO_3^{2-}	Hydrogen-sulphite	HSO_3^-	Phosphate	PO_4^{3-}
Chromate	CrO_4^{2-}	Hydroxide	OH^-	Silicate	SiO_3^{2-}
Dichromate	$Cr_2O_7^{2-}$	Methanoate	$HCOO^-$	Sulphate	SO_4^{2-}
Ethanoate	CH_3COO^-	Nitrate	NO_3^-	Sulphite	SO_3^{2-}
Hydrogen-carbonate	HCO_3^-	Nitrite	NO_2^-	Thiosulphate	$S_2O_3^{2-}$

Guide to Solubility in Water

The table below is intended to give some indication of the solubility of selected substances in water at room temperature. A simple coding system is used in preference to quoting actual data. Details of the code are given below.

	Bromide	Carbonate	Chloride	Fluoride	Iodide	Nitrate	Oxide (Hydroxide)	Sulphate	Sulphite
Aluminium	a	—	a	b	a	a	i	a	—
Ammonium	a	a	a	a	a	a	a	a	a
Barium	a	c	a	b	a	a	a	—	c
Calcium	a	c	a	—	a	a	b	b	—
Copper(II)	a	i	a	a	—	a	i	a	—
Iron(III)	a	—	a	—	—	a	i	a	—
Lead(II)	b	i	b	c	c	a	c	c	i
Magnesium	a	c	a	c	a	a	i	a	—
Potassium	a	a	a	a	a	a	a	a	a
Silver	i	c	i	a	i	a	c	b	a
Sodium	a	a	a	a	a	a	a	a	a
Zinc	a	c	a	a	a	a	—	a	b

Notes

The letter "a" is to be taken to indicate a solubility greater than 10 g l^{-1}.
The letter "b" is to be taken to indicate a solubility between 1 g l^{-1} and 10 g l^{-1}.
The letter "c" is to be taken to indicate a solubility between 0·01 g l^{-1} and 1 g l^{-1}.
The letter "i" is to be taken to indicate a solubility less than 0·01 g l^{-1}.

Appendix

Periodic Table

Legend:
- ATOMIC NUMBER
- SYMBOL
- ELECTRON ARRANGEMENT
- Name

1 H 1 Hydrogen																	2 He 2 Helium
3 Li 2,1 Lithium	4 Be 2,2 Beryllium											5 B 2,3 Boron	6 C 2,4 Carbon	7 N 2,5 Nitrogen	8 O 2,6 Oxygen	9 F 2,7 Fluorine	10 Ne 2,8 Neon
11 Na 2,8,1 Sodium	12 Mg 2,8,2 Magnesium											13 Al 2,8,3 Aluminum	14 Si 2,8,4 Silicon	15 P 2,8,5 Phosphorus	16 S 2,8,6 Sulphur	17 Cl 2,8,7 Chlorine	18 Ar 2,8,8 Argon
19 K 2,8,8,1 Potassium	20 Ca 2,8,8,2 Calcium	21 Sc 2,8,9,2 Scandium	22 Ti 2,8,10,2 Titanium	23 V 2,8,11,2 Vanadium	24 Cr 2,8,13,1 Chromium	25 Mn 2,8,13,2 Manganese	26 Fe 2,8,14,2 Iron	27 Co 2,8,15,2 Cobalt	28 Ni 2,8,16,2 Nickel	29 Cu 2,8,18,1 Copper	30 Zn 2,8,18,2 Zinc	31 Ga 2,8,18,3 Gallium	32 Ge 2,8,18,4 Germanium	33 As 2,8,18,5 Arsenic	34 Se 2,8,18,6 Selenium	35 Br 2,8,18,7 Bromine	36 Kr 2,8,18,8 Krypton
37 Rb 2,8,18,8,1 Rubidium	38 Sr 2,8,18,8,2 Strontium	39 Y 2,8,18,9,2 Yttrium	40 Zr 2,8,18,10,2 Zirconium	41 Nb 2,8,18,11,2 Niobium	42 Mo 2,8,18,13,1 Molybdenum	43 Tc 2,8,18,14,1 Technetium	44 Ru 2,8,18,15,1 Ruthenium	45 Rh 2,8,18,16,1 Rhodium	46 Pd 2,8,18,18,0 Palladium	47 Ag 2,8,18,18,1 Silver	48 Cd 2,8,18,18,2 Cadmium	49 In 2,8,18,18,3 Indium	50 Sn 2,8,18,18,4 Tin	51 Sb 2,8,18,18,5 Antimony	52 Te 2,8,18,18,6 Tellurium	53 I 2,8,18,18,7 Iodine	54 Xe 2,8,18,18,8 Xenon
55 Cs 2,8,18,18,8,1 Caesium	56 Ba 2,8,18,18,8,2 Barium	57 La 2,8,18,18,9,2 Lanthanum	72 Hf 2,8,18,32,10,2 Hafnium	73 Ta 2,8,18,32,11,2 Tantalum	74 W 2,8,18,32,12,2 Tungsten	75 Re 2,8,18,32,13,2 Rhenium	76 Os 2,8,18,32,14,2 Osmium	77 Ir 2,8,18,32,17,0 Iridium	78 Pt 2,8,18,32,17,1 Platinum	79 Au 2,8,18,32,18,1 Gold	80 Hg 2,8,18,32,18,2 Mercury	81 Tl 2,8,18,32,18,3 Thallium	82 Pb 2,8,18,32,18,4 Lead	83 Bi 2,8,18,32,18,5 Bismuth	84 Po 2,8,18,32,18,6 Polonium	85 At 2,8,18,32,18,7 Astatine	86 Rn 2,8,18,32,18,8 Radon
87 Fr 2,8,18,32,18,8,1 Francium	88 Ra 2,8,18,32,18,8,2 Radium	89 Ac 2,8,18,32,18,9,2 Actinium															

LANTHANIDES

57 La 2,8,18,18,9,2 Lanthanum	58 Ce 2,8,18,20,8,2 Cerium	59 Pr 2,8,18,21,8,2 Praseodymium	60 Nd 2,8,18,22,8,2 Neodymium	61 Pm 2,8,18,23,8,2 Promethium	62 Sm 2,8,18,24,8,2 Samarium	63 Eu 2,8,18,25,8,2 Europium	64 Gd 2,8,18,25,9,2 Gadolinium	65 Tb 2,8,18,27,8,2 Terbium	66 Dy 2,8,18,28,8,2 Dysprosium	67 Ho 2,8,18,29,8,2 Holmium	68 Er 2,8,18,30,8,2 Erbium	69 Tm 2,8,18,31,8,2 Thulium	70 Yb 2,8,18,32,8,2 Ytterbium	71 Lu 2,8,18,32,9,2 Lutetium

ACTINIDES

89 Ac 2,8,18,32,18,9,2 Actinium	90 Th 2,8,18,32,18,10,2 Thorium	91 Pa 2,8,18,32,20,9,2 Protactinium	92 U 2,8,18,32,21,9,2 Uranium	93 Np 2,8,18,32,22,9,2 Neptunium	94 Pu 2,8,18,32,24,8,2 Plutonium	95 Am 2,8,18,32,25,8,2 Americium	96 Cm 2,8,18,32,25,9,2 Curium	97 Bk 2,8,18,32,26,9,2 Berkelium	98 Cf 2,8,18,32,28,8,2 Californium	99 Es 2,8,18,32,29,8,2 Einsteinium	100 Fm 2,8,18,32,30,8,2 Fermium	101 Md 2,8,18,32,31,8,2 Mendelevium	102 No 2,8,18,32,32,8,2 Nobelium	103 Lr 2,8,18,32,32,9,2 Lawrencium

Index

Acetic acid – *see* ethanoic acid
Acidic oxides, 55
Acids, 55–57
   formula, 37
   organic, naming, 125
   pH of, 57
   reactions, 56, 64
   strength and concentration, 59
   strong and weak, 58–60
Activity series, 39
Addition of bromine, 112
   polymerisation, 112, 133
Alkalis, 55, 57, 64
   pH, 58
Alkanes, 104
   reactions, 106
   combustion, 109
   isomers, 105, 106
Aluminium, anodising, 52
   corrosion, 52
Amino acids, 130
   condensation, 131
Ammonia, 90
   alkaline nature, 90
   and copper (II) oxide, 94
   and water, 90
   burning in oxygen, 93
   catalytic oxidation, 94
   manufacture, 92
   oxidation, 93, 94
   solution conductivity, 90
   weak base, 90
Ammonium
   chloride, 90
   hydrogen phosphate, 98
   ions, 90
   nitrate, 98
   sulphate, 98
Amylase, 118, 123
Anodising, 52
Atom representation, 2, 6
   structure, 2, 4
Atomic mass, 6
   mass units, 3
   number, 4
   weight, 6
Atoms combining, 12

Bakelite, 136
Bases, 55, 56, 57, 64
   strong and weak, 60
Basic oxides, *see* bases
Bond breaking, 109
   making, 109
Brewing, 123
Brownian movement, 1
Burning alkanes, 109
   carbon, 103
   carbon monoxide, 104
   petrol, 103
   plastics, 114, 136
Butane, formula and structure, 105

Calcium phosphate, 99
Carbohydrates, 116
   as energy sources, 118
   classes of, 117
Carbon, 101
   and air, 104
   as a fuel, 102
   burning, 103
   cycle, 120
   dioxide, 103
   forming chains, 104
   polymorphs, 101
Carbon monoxide, 103
   as a fuel, 104
   poisonous nature, 104
Carbonates, 57, 64
Catalytic cracking, 110
Cathodic protection of iron, 50
Chlorophyll, 119
Chromatography of amino acids, 131
   of carbohydrates, 121
   of hydrocarbons, 107

Index

Combination of atoms, 12
Combustion of alkanes, 109
  carbon, 103
  carbon monoxide, 104
  oil fractions, 109
  petrol, 103
Concentration of solutions, 31
  determination of, 71
Condensation of amino acids, 131
  of acids and alcohols, 125
  of glucose, 118
  polymerisation, 118, 134
  to esters, 125
  to fats and oils, 127
  to starch, 118
Conductivity, and neutralisation, 67–71
  effect of concentration, 65
  graphs, 67–70
  mobility of ions, 65
Conservations of sewage, 99
Contact Process, 84
Corrosion of aluminium – see aluminium
  of iron – see iron
Covalent bond, 13, 14, 15–18
  formulae, 28

Detergents, 130
Diamond, 101
Diastase – see amylase
Diffusion, 1
Dinitrogen tetroxide, 98
Disaccharides, 117, 121
Discharge in electrolysis, 74–77
Displacement of metals, 40
  and reacting masses, 41

Electrochemical series, 44
Electrolysis, 74–77
  copper (II) chloride, 9, 74
  lead bromide, 9
  order of discharge of ions, 77
  sodium halides, 75
  sulphuric acid, 76
  water, 76
Electrolytes, 9
Electron arrangement, 7, 11
  attraction, 14
  clouds – see orbitals
  energy levels, 6, 11
Electron transfer, 41
  and the sulphate ion, 87
  and the nitrate ion, 96–98
  in redox reactions, 47
  and the sulphite ion, 80, 81

Electroplating, 51
Electrovalent bond – see ionic bond
Enzymes, 118, 123, 131
Equations, balanced, 34
  in calculations, 34, 35
  ion-electron, 41, 45–47
  ionic, 36
  state, 36
  unbalanced, 34
Equilibrium of water, 61
Ethane, formula and structure, 105
Ethanoic acid, 124, 125
  properties, 125
  structure, 125
  weak acid, 125
Ethanol, 122
  from glucose, 123
  oxidation, 124
  properties, 124
  structure, 123
Ethene, 111
  polymerisation, 112, 133
Esters, hydrolysis, 126
  making, 125
  naming, 126
  uses, 126

Fats, 127
  as energy sources, 127
  into soap, 128
Fermentation of glucose
  to ethanol, 123
Ferroxyl indicator, 47
Formula, by experiment, 30
  covalent, 15, 28
  ionic, 27
  mass (weight), 29
  writing, 28
Fractional distillation, 108

Galvanising, 47, 51
Gas chromatography, 107
Glucose, 117
  fermentation, 123
Glycerol, 127
Glycogen, 117, 120
Gram formula mass (weight), 29, 30
Graphite, 101

Haber Process, 92, 139
Homologous series, 104
Hydrocarbons, 104
  as fuels, 109
  burning, 106

composition, 106
separation, 107
thermal cracking, 110
Hydrogen ion, 55–57
  mobility, 66
Hydrogen sulphate ion, 86
Hydrolysis, 117
  of esters, 126
  of maltose, 122
  of proteins, 131
  of starch, 117
  of sucrose, 121
Hydroxide ion, 57
  mobility, 66

Indicators, ferroxyl, 47
  methyl orange, 68
  pH, 57, 58
Ion-electron equations, 41, 45
Ionic bond, 13, 18
  compounds, 19
  formulae, 25, 27
  shapes of compounds, 20
Ions, 10, 22–24
  colour of, 22
  ease of formation, 43
  large, 23
  mobility, 65
  spectator, 36, 69, 71
Iron, corrosion, 47–50
  galvanising (zinc plating), 47, 51
  protection, 47, 50, 51
  tin plating, 47, 51
Isomers, 105
Isotopes, 6

Macromolecule, 101, 133
Maltase, 123
Maltose hydrolysis, 122
Margarine, 128
Mass number, 4–6
  determination of, 4
Mass spectrometer, 4
Metallic conductors, 9
Metal oxide reduction, 42
Metals and acid, 40, 56, 64
  and oxygen, 39
  and water, 39
  displacement, 40–42
Methane, formula and structure, 105
Mobility of ions, 65
Molarity, 31
Mole, 30
Molecules, 10, 15–18

shape, 17, 18
structure, 15–17
Monosaccharides, 117

Natural addition polymers, 133
  condensation polymers, 134
Natural gas, 108
Neutralization, 56, 57
  and conductivity, 68–71
  heat of, 71
Nitrate ion tests, 95
Nitric acid, 95
  oxidising action, 96–98
  pH, 96
  properties, 96–98
Nitric oxide – *see* nitrogen monoxide
Nitrogen, 89
  and hydrogen, 91
  compounds, 98
  dioxide, 90, 95, 97, 98
  fertilisers, 98
  monoxide, 95, 97
  sparking with air, 89
Non-conductors, 9, 10
Nylon, 134

Oil, 108
  fractional distillation, 108
  vegetable, 127
Oleic acid, 127
Oleum, 84, 129
Orbitals, 11
  half-filled, 12
  tetrahedral arrangement, 11
Ostwald Process, 94, 140
Oxidation, 45
Oxides, reduction, 42

Paper chromatography, 121, 131
Pepsin, 131
Peptide link, 131
Percentage composition, 29
pH of acids, 57, 58
  of alkalis, 58
  scale, 57, 58
Phosphates, 100
Photosynthesis, 119
Polymerisation, 112, 118, 133
  addition, 112, 133
  condensation, 134
  to nylon, 134
  to polyproplene, 113
  to polystyrene, 114

Index

Polymerisation (contd.)
  to polythene, 112
  to polyvinyl chloride (PVC), 113
  to rayon, 134
  to starch, 118
  to teflon (PTFE), 113
  to urea-formaldehyde, 136
Polysaccharides, 117
Potassium hexacyanoferrate (III), 46, 47
Precipitation, 65, 70, 138
Proteins, 130
  composition, 130
  digestion, 131
  from amino acids, 131
  hydrolysis, 131
Ptyalin — see amylase

Rayon, 134
Redox, 45
Reduction, 45
  in electrolysis, 46
Reforming, 110, 111
Respiration, 119
  in animals, 120
Rubber, 133

Saccharides, tests, 117
Sacrificial protection, 50
Salts, naming, 63
  preparation, 64, 138
  solubility, 63
Saturated hydrocarbons, 106
  reactions, 106, 107, 112
Sewage conservation, 99
Silicones, 137
Soap, and hard water, 129
  cleansing (detergent) action, 128
  making, 128
Soapless detergents, 129
Solubility, 63
Spectator ions, 36, 69, 71
Starch, 117
  from glucose, 118
  test, 117
State equations, 36
Stearic acid, 127
Substitution reactions, 106, 112
Sucrose hydrolysis, 121
Sulphate ion, 84, 87
  test for, 80
Sulphite ion, 79

  reactions, 80
  reducing action, 80
  uses, 82
Sulphur dioxide, 78
  making, 78, 82
  pollution, 79
  properties, 79
  reducing action, 80
  tests, 78
Sulphuric acid
  and metals, 87
  attraction for water, 86
  ionisation, 84, 87
  manufacture, 84
  oxidising action, 87
  preparation, 82
  properties, 84–87
  uses, 87
Sulphurous acid, 79
  making, 83, 84
Sulphur trioxide, 82
Superphosphate, 100

Test for hydroxide ion, 57, 58
  iron(II) ion, 46, 47
  nitrate ion, 95
  starch, 117
  sugars, 117
  sulphate ion, 80
  sulphur dioxide, 78
  unsaturated hydrocarbons, 112
Thermal cracking, 110
Thermoplastics, 135
Thermosets, 135
Tin plating, 47, 51

Unsaturated hydrocarbons, 111
  and bromine, 112
  in polymerisation, 112, 133
  test for, 112
Urea formaldehyde, 136

Valency, 25
Vulcanising, 133

Water, equilibrium, 61
  ionisation, 60
  pH, 58

Zymase, 123